STUDIES IN MODEL THEORY

Studies in Mathematics

The Mathematical Association of America

Volume 1: STUDIES IN MODERN ANALYSIS
edited by R. C. Buck

Volume 2: STUDIES IN MODERN ALGEBRA
edited by A. A. Albert

Volume 3: STUDIES IN REAL AND COMPLEX ANALYSIS
edited by I. I. Hirschman, Jr.

Volume 4: STUDIES IN GLOBAL GEOMETRY AND ANALYSIS
edited by S. S. Chern

Volume 5: STUDIES IN MODERN TOPOLOGY
edited by P. J. Hilton

Volume 6: STUDIES IN NUMBER THEORY
edited by W. J. LeVeque

Volume 7: STUDIES IN APPLIED MATHEMATICS
edited by A. H. Taub

Volume 8: STUDIES IN MODEL THEORY
edited by M. D. Morley

Jon Barwise
University of Wisconsin, Madison

A. R. Bernstein
University of Maryland, College Park

C. C. Chang
University of California, Los Angeles

H. J. Keisler
University of Wisconsin, Madison

M. D. Morley
Cornell University

Abraham Robinson
Yale University

J. H. Silver
University of California, Berkeley

Studies in Mathematics

Volume 8

STUDIES IN MODEL THEORY

M. D. Morley, editor
Cornell University

Published and distributed by
The Mathematical Association of America

Library of Congress Catalog Card Number 73-86564

Complete Set ISBN 0-88385-100-8
Vol. 8 ISBN 0-88385-108-3

Printed in the United States of America

Current printing (last digit):

10 9 8 7 6 5 4 3 2 1

CONTENTS

STUDIES IN MODEL THEORY

INTRODUCTION AND GUIDE TO THE READER

A mathematical field involves the study of two different kinds of objects. For instance, in group theory, both groups and statements about groups are studied. Model theory may be described as the study of the relationships between classes of mathematical objects and classes of statements about them. The purpose of this study is to increase both our knowledge of the structure of classes of mathematical objects and of the languages in which statements about these classes are formulated.

This book is intended to give a sample of the kinds of methods used and results obtained in model theory. The reader should be mathematically literate but need not be knowledgeable about formal logic. This introduction and Appendix I cover the few facts of formal logic which are necessary for the reader to know.

The results of model theory naturally depend on the class of admissible statements, that is, the formal language in which they are phrased. There are many possibilities, but there is one sort of language more important than others. This is the lower (or first order) predicate calculus. Such a language will contain symbols corresponding to finitary relations and finitary operations on the mathematical objects under study. It will have connectives ("and", "or", "not", etc.) and quantifiers $\exists x$ ("there exists an x such that...") and $\forall x$ ("for all x . . ."). Finally, it will have variables ranging over the elements of the object under consideration. The reason it is called the first order predicate calculus is that it does not contain other types of variables; for example, it has no variables ranging over subsets of the object under study.

Many of the notions of algebra are first order (i.e., expressible in a first order language). The axioms for groups, rings, fields and linear orderings are all first order. Topological notions, in general, are not first order. Peano's postulates for the natural numbers are not first order since the induction axiom ("For every set Y, if $0 \in Y$ and $n \in Y$ implies $n + 1 \in Y$, then every $n \in Y$.") contains a variable Y ranging over sets. Thus, notions whose definition includes variables ranging over natural numbers are generally not first order. The notions of field and of ordered field are first order but the notion of Archimedean ordered field is not.

Despite these restrictions, the lower predicate calculus is the starting point for model theory because it has some very nice properties. The most important of these is the compactness property: a set of sentences is consistent if every finite subset is. (Consistent means there exists some mathematical object in which every sentence of the set is true.) Most results in model theory involve some use, possibly disguised, of the compactness property. Another important property is that of Löwenheim-Skolem: A countable consistent set of sentences is valid in some mathematical object which has only a countable set of elements.

Appendix I contains a more detailed list of definitions and results from the predicate calculus. The reader unfamiliar with the predicate calculus should consult this list, but he should not permit technicalities to cloud his view of the basic facts about the predicate calculus: its variables range only over individuals and it has the compactness property.

The reader completely unfamiliar with model theory is advised to begin with the article by A. Robinson. In it, he describes very clearly not only how model theory increases our understanding of fields and their algebraic closures but leads to the generalization of the notion of algebraic closure to other objects.

The notion of saturated model was also inspired (partly) by the notion of algebraic closure. It is a generalization of the η_α linearly ordered sets of Hausdorf. C. C. Chang's article describes saturated models and several of their remarkably diverse uses.

Model theorists are no more limited to statements that can be physically written down than geometers are limited to Euclidean 3-space. J. Barwise considers a certain class of infinitely long formulas and relates them to properties of certain sets of mappings. The relation of these to the invariants of countable Abelian groups is particularly striking.

As mentioned earlier, topological notions are not expressible in first order language. In an attempt to circumvent this difficulty, we can introduce a first order language with different types of variables, the types corresponding to individuals, sets, sets of sets, etc., and to consider the relation $x \in y$ as just another binary relation. But these devices will not completely capture the intended meaning and the axioms for integers, real numbers, etc., will have "non-standard" models non-isomorphic to the intended ones. Bernstein's article, which presents the solution of an invariant subspace problem of analysis, shows how nonstandard models can be used to solve problems in standard models.

Another way to reduce mathematics to first order logic is to observe that:

(i) all mathematics can be reduced to set theory and

(ii) the intuitive content of set theory is expressible in a set of first order axioms about the binary relation $\in$.

Jack Silver shows how model theoretical methods can be used to obtain quite surprising results in set theory. He uses the most popular set of axioms for set theory, Zermelo-Fraenkel (Z.F.).

It is not necessary to know the exact form of the axioms of Z.F. to understand Silver's article. The reader need only know that the axioms are first order and have faith that all the usual assumptions about sets follows from them. For you of little faith, the axioms of Z.F. are stated in Appendix III.

The growth of model theory led not only to applications of logic to other branches of mathematics but also to applications of increasingly sophisticated mathematical methods to logic. An example of this is the ultrapower construction. Ultrapowers, described in Appendix II, appear in Bernstein's article as well as Silver's.

The article of H. J. Keisler is an excellent example of this increased sophistication. It introduces a lovely notion which unifies several concepts. Even the most knowledgeable will find new ideas here.

As editor, I wish to express my appreciation to the six authors of the articles. They were chosen not only because each had made a significant contribution to model theory but also because of their expository gifts. Each has accomplished the nearly inconsistent task of writing an article useful both to those ignorant of model theory and to those expert in it. The editor is amazed and grateful.

MICHAEL D. MORLEY

BACK AND FORTH THROUGH INFINITARY LOGIC

Jon Barwise

In model theory, as in every mature branch of mathematics, certain techniques have emerged as basic. One of these is the so called "back and forth argument". In this paper we introduce the reader to the simplest kind of back and forth argument, the "one at a time" version. In particular, we draw the reader's attention to a kind of isomorphism relation $\cong_p$ which is generally overlooked, but which, from a foundational point of view, is quite natural. We show how $\cong_p$ is related to infinitary logic in sections 3 and 4, and two of its algebraic applications in sections 5 and 6. Its foundational significance is discussed in section 7, where some historical and bibliographical material can also be found. Sections 6 and 7 can be read following section 3 if desired.

1. CANTOR'S THEOREM

The father of the back and forth argument is Georg Cantor. He used it to prove:

THEOREM 1: *Any countable dense linearly ordered set without endpoints is isomorphic to the set of rational numbers with the natural ordering.*

Or, equivalently, *any two countable dense linearly ordered sets without endpoints are isomorphic.* The proof breaks naturally into two parts.

Let $\mathfrak{A} = \langle A, <^{\mathfrak{A}} \rangle$ and $\mathfrak{B} = \langle B, <^{\mathfrak{B}} \rangle$ be dense linearly ordered sets without endpoints, countable or not, and define the following set I of partial isomorphisms:

$$I = \{f : f \text{ is an isomorphism of a finite subordering of } \mathfrak{A} \text{ onto a finite subordering of } \mathfrak{B}\}.$$

The first step in the proof is to show that I has the

Back and Forth Property: For every $f \in I$ and $a \in A$ (resp. $b \in B$) there is a $g \in I$ with $f \subseteq g$ and $a \in \text{dom}(g)$ (resp. $b \in \text{rng}(g)$).

(By $f \subseteq g$ we mean that g is an extension of f, dom (g) and rng (g) are the domain and range of g respectively.) For example, if dom $(f) = \{x_1 < x_2 < \cdots < x_n\}$, rng $(f) = \{y_1 < y_2 < \cdots < y_n\}$ and $x_2 < a < x_3$ then we can choose any b between y_2 and y_3 for $g(a)$.

Now, if A and B are countable, say $A = \{a_1, a_2, \cdots\}$ and $B = \{b_1, b_2, \cdots\}$ then we can use the back and forth property of I to construct an isomorphism of $\mathfrak{A}$ and $\mathfrak{B}$. Let $f_1 = \{\langle a_1, b_1 \rangle\}$ be the function which maps a_1 to b_1, so that $f_1 \in I$, and define a sequence

$$f_1 \subseteq f_2 \subseteq \cdots \subseteq f_n \subseteq \cdots$$

of functions as follows:

$$f_{2n} = \text{some function } g \in I \text{ with } f_{2n-1} \subseteq g \text{ and } a_n \in \text{dom}(g)$$

$$f_{2n+1} = \text{some function } g \in I \text{ with } f_{2n} \subseteq g \text{ and } b_n \in \text{rng}(g).$$

The limit $f = \bigcup_n f_n$ of these functions clearly has domain all of A, range all of B, and preserves $<^{\mathfrak{A}}$ and $<^{\mathfrak{B}}$, that is, f is an isomorphism of $\mathfrak{A}$ onto $\mathfrak{B}$. ⊣

There is a feeling that the second half of this proof has very little to do with linear orderings. We can turn this feeling into a theorem as follows:

Given structures $\mathfrak{A}$ and $\mathfrak{B}$ of the same similarity type (for example, if $\mathfrak{A} = \langle A, R, F\rangle$ with R a binary relation and F a 3-ary function on A, then $\mathfrak{B}$ has the same similarity type if $\mathfrak{B} = \langle B, S, G\rangle$ with S a binary relation and G a 3-ary function on B) we write

$$I: \mathfrak{A} \cong_p \mathfrak{B},$$

if I is a nonempty set of isomorphisms of substructures of $\mathfrak{A}$ onto substructures of $\mathfrak{B}$ and I has the back and forth property. We say that $\mathfrak{A}$ and $\mathfrak{B}$ are partially isomorphic and write

$$\mathfrak{A} \cong_p \mathfrak{B},$$

if there is an I such that $I: \mathfrak{A} \cong_p \mathfrak{B}$. We write $f: \mathfrak{A} \cong \mathfrak{B}$ to mean that f is an isomorphism of $\mathfrak{A}$ onto $\mathfrak{B}$ and $\mathfrak{A} \cong \mathfrak{B}$ if $\mathfrak{A}$ and $\mathfrak{B}$ are isomorphic.

THEOREM 2: *If $\mathfrak{A}$ and $\mathfrak{B}$ are countable, or countably generated,*[1] *structures then $\mathfrak{A} \cong \mathfrak{B}$ iff $\mathfrak{A} \cong_p \mathfrak{B}$. In fact, if $I: \mathfrak{A} \cong_p \mathfrak{B}$ and $f_0 \in I$ then f_0 can be extended to an isomorphism $f: \mathfrak{A} \cong \mathfrak{B}$.*

Proof: If $f: \mathfrak{A} \cong \mathfrak{B}$ then setting $I = \{f\}$ we have

$$I: \mathfrak{A} \cong_p \mathfrak{B}.$$

To prove the converse, let $\mathfrak{A}$ be generated by $\{a_1, a_2, \cdots\}$, $\mathfrak{B}$ by $\{b_1, b_2, \cdots\}$ and let $I: \mathfrak{A} \cong_p \mathfrak{B}$. Then $I \neq \emptyset$ so let $f_0 \in I$. Now proceed as in the proof of Theorem 1 to adjoin elements a_n and b_n in turn. ⊣

Before leaving this section let us agree on what we have been meaning by the word "substructure". For example, let $\mathfrak{A} = \langle A, R, F\rangle$ be as in the parenthetical remark above. Normal usage in model theory would require a substructure $\mathfrak{A}_0$ of $\mathfrak{A}$ to be of the form $\langle A_0, R \cap A_0^2, F \restriction A_0^3\rangle$ where A_0 is a nonempty subset of A

1. All results stated here for countable structures hold just as well for countably generated structures. This is of importance if one is interested in, for example, modules over an uncountable ring, as in [15].

closed under F. It is more convenient for us to remove the assumption that A_0 is nonempty, so that the empty function can be considered an isomorphism of a substructure of $\mathfrak{A}$ onto a substructure of $\mathfrak{B}$. Notice that this in no way harms the proof of Theorem 2. Of course if $\mathfrak{A}$ is $\langle A, R, F, a\rangle$ with $a \in A$ then a substructure $\mathfrak{A}_0 = \langle A_0, R \cap A_0^2, F \restriction A_0^3, a\rangle$ must have $a \in A_0$, so that in this case $a \in \text{dom}(f)$ for any partial isomorphism f.

Examples.

1. Let $\mathfrak{A} = \langle A, <^{\mathfrak{A}}\rangle$ and $\mathfrak{B} = \langle B, <^{\mathfrak{B}}\rangle$ be dense linearly ordered sets with first elements a and b respectively but without last elements. The set I used in Theorem 1 no longer has the back and forth property, but

$$I_0 = \{f \in I: a \in \text{dom}(f) \text{ and } f(a) = b\}$$

does, so $I_0: \mathfrak{A} \cong_p \mathfrak{B}$.

2. A Boolean algebra $\mathfrak{B} = \langle B, \cap, \cup, \sim, 0, 1\rangle$ is atomless if for every $b \neq 0$ there is a c, $0 \neq c \neq b$, with $c \leqq b$. If $\mathfrak{B}_0$ and $\mathfrak{B}_1$ are any two atomless Boolean algebras, then

$$\mathfrak{B}_0 \cong_p \mathfrak{B}_1.$$

3. The first two examples show that all countable models of certain first order theories are isomorphic, if one uses Theorem 2. We give a different sort of example here. Let us call a subgroup G_0 of an Abelian group G *thin in* G if G_0 is finitely generated and $G = G_0 \oplus G_1$ with G_1 isomorphic to G. For each j in an infinite index set J let G_j be an infinite cyclic group generated by, say, x_j. Let $G = \sum_j G_j$ be the direct sum and $H = \prod_J G_j$ the direct product of the G_j. Let

$$I = \{f: f \text{ is an isomorphism of a thin subgroup of } G \text{ onto a thin subgroup of } H\}.$$

Then

$$I: G \cong_p H.$$

To show this, define $\|h\|$ for $h = \langle n_j x_j: j \in J\rangle$ in H, to be the smallest nonzero value of $|n_j|$. Prove by induction on $\|h\|$ that h is an element of a thin subgroup of H. (For more help see Fuchs [5] p. 168.) Note that G and H may have different cardinality.

2. THE LANGUAGE $L_{\infty\omega}$

We assume that the reader understands the notion of a first order language L with identity, explained elsewhere in this book. The logical symbols of L we take to be $\neg$ (not), $\bigwedge$ (and), $\bigvee$ (or), $\forall$ (for all), $\exists$ (there exists), $=$ (equality) and variables $v_0, v_1, \cdots, v_n, \cdots$.

The language $L_{\infty\omega}$ has the same symbols as L, except that we add variables v_α for all ordinals α and allow them to occur in atomic formulas. The class of *formulas* of $L_{\infty\omega}$ is the smallest class Y containing the atomic formulas of L and closed under (1)–(3).

(1) If $\varphi \in Y$ then $(\neg\varphi) \in Y$.
(2) If $\varphi \in Y$ then $(\forall v_\alpha \varphi)$ and $(\exists v_\alpha \varphi)$ are in Y.
(3) If Φ is a subset of Y, finite or infinite, then the conjunction $\bigwedge\Phi$ and disjunction $\bigvee\Phi$ are in Y.

The notions of free and bound variable are defined in a manner parallel to that for L. We use $\varphi(x_1 \cdots x_n)$ to denote a formula whose free variables are included in the set $\{x_1 \cdots x_n\}$. The notation

$$\mathfrak{A} \models \varphi[a_1 \cdots a_n],$$

means that $\mathfrak{A}$ is a structure for L and φ is true in $\mathfrak{A}$ when x_i is interpreted as a name of a_i for $i = 1, \cdots, n$. A *sentence* φ is a formula with no free variables and $\mathfrak{A} \models \varphi$ means that φ is true in $\mathfrak{A}$, also read, $\mathfrak{A}$ is a model of φ.

The use of infinite expressions allows us to express many important notions not expressible in L. For the examples below, suppose that L is the language for Abelian groups with a binary function symbol $+$ and a constant symbol $\mathbf{0}$. For any term t of L we define $n \cdot t$ by recursion for integers $n \geq 0$:

$0 \cdot t$ is the term $\mathbf{0}$,

$(n + 1) \cdot t$ is the term $(n \cdot t + t)$.

Examples.

1. If G is an Abelian group, then G is a torsion group iff G is a model of

$$\forall x \bigvee \{n \cdot x = \mathbf{0}: 0 < n < \omega\}.$$

This infinite sentence is usually written more informally as

$$\forall x \bigvee_{0<n<\omega} n \cdot x = 0.$$

2. To say that y is in the subgroup generated by $x_1 \cdots x_k$ causes a little trouble since we have not allowed a function symbol for subtraction. However, $y = x - z$ iff $x = y + z$, so the desired predicate of $y, x_1 \cdots x_n$ can be expressed by $\bigvee\Phi$, where Φ is the set of formulas of the form

$$n_1 \cdot x_1 + \cdots + n_k \cdot x_k = y + m_1 \cdot x_1 + \cdots + m_k \cdot x_k,$$

where $0 \leqq n_i < \omega$ and $0 \leqq m_i < \omega$ for $i = 1 \cdots k$.

3. If we call the formula of the above example $\varphi_k(x_1 \cdots x_k, y)$ then the sentence

$$\bigvee_{0<k<\omega} \exists x_1 \cdots \exists x_k \forall y \varphi_k(x_1 \cdots x_k, y),$$

is true in an Abelian group G iff G is finitely generated.

There are many measures of the complexity of a formula φ of $L_{\infty\omega}$. In this article, we need only two of the most simple-minded, $qr(\varphi)$ and $|\varphi|$.

The quantifier rank of a formula φ, $qr(\varphi)$, is an ordinal number[2] defined by induction on formulas as follows:

$$qr(\varphi) = 0, \quad \text{if } \varphi \text{ is atomic,}$$

$$qr(\neg\varphi) = qr(\varphi),$$

$$qr(\forall v_\alpha \varphi) = qr(\exists v_\alpha \varphi) = qr(\varphi) + 1,$$

$$qr(\bigvee\Phi) = qr(\bigwedge\Phi) = \sup\{qr(\varphi): \varphi \in \Phi\}.$$

2. We refer the reader to Halmos [6] for the rudiments of ordinal and cardinal arithmetic.

The cardinality of φ, $|\varphi|$, is by definition the cardinality of the set sub (φ) of subformulas of φ, where:

$$\text{sub}(\varphi) = \{\varphi\} \quad \text{if } \varphi \text{ is atomic,}$$

$$\text{sub}(\neg\varphi) = \text{sub}(\varphi) \cup \{\neg\varphi\},$$

$$\text{sub}(\varphi) = \text{sub}(\psi) \cup \{\varphi\} \quad \text{if } \varphi \text{ is } \forall v_\alpha \psi \text{ or } \exists v_\alpha \psi,$$

$$\text{sub}(\varphi) = \{\varphi\} \cup \bigcup_{\psi \in \Phi} \text{sub}(\psi) \quad \text{if } \varphi \text{ is } \bigvee\Phi \text{ or } \bigwedge\Phi.$$

Notice also that if φ has only a finite number of distinct free variables then so does any $\psi \in \text{sub}(\varphi)$. In particular, if ψ is a subformula of some sentence φ then ψ has at most a finite number of free variables. This accounts for our interest only in formulas with a finite number of free variables.

Given structures $\mathfrak{A}$ and $\mathfrak{B}$ for L, we write

$$\mathfrak{A} \equiv_{\infty\omega} \mathfrak{B},$$

if $\mathfrak{A}$ and $\mathfrak{B}$ are models of the same sentences of $L_{\infty\omega}$. For an ordinal α we write

$$\mathfrak{A} \equiv^\alpha_{\infty\omega} \mathfrak{B},$$

if for all sentences φ of $L_{\infty\omega}$ with $qr(\varphi) \leqq \alpha$ we have $\mathfrak{A} \models \varphi$ iff $\mathfrak{B} \models \varphi$. Thus $\mathfrak{A} \equiv_{\infty\omega} \mathfrak{B}$ iff $\mathfrak{A} \equiv^\alpha_{\infty\omega} \mathfrak{B}$ for all α.

If $\mathfrak{A} = \langle A, \cdots \rangle$ is a structure for L and $a_1 \cdots a_n \in A$ then we let $(\mathfrak{A}, a_1 \cdots a_n)$ be the structure $\mathfrak{A} = \langle A, \cdots, a_1 \cdots a_n \rangle$. This is a structure for the language L' obtained by adding names $\boldsymbol{a}_1 \cdots \boldsymbol{a}_n$ to L. To say that $(\mathfrak{A}, a_1 \cdots a_n) \equiv_{\infty\omega} (\mathfrak{B}, b_1 \cdots b_n)$ means that the two structures satisfy the same sentences of $L'_{\infty\omega}$. This can also be expressed by saying that for all formulas $\varphi(x_1 \cdots x_n)$ of $L_{\infty\omega}$,

$$\mathfrak{A} \models \varphi[a_1 \cdots a_n] \quad \text{iff} \quad \mathfrak{B} \models \varphi[b_1 \cdots b_n].$$

There are corresponding notions of elementary substructure. Given $\mathfrak{A} \subseteq \mathfrak{B}$, we write

$$\mathfrak{A} \prec_{\infty\omega} \mathfrak{B},$$

if for any formula $\varphi(x_1 \cdots x_n)$ and any $a_1 \cdots a_n \in A$,

$$\mathfrak{A} \models \varphi[a_1 \cdots a_n] \quad \text{iff} \quad \mathfrak{B} \models \varphi[a_1 \cdots a_n],$$

and we write

$$\mathfrak{A} \prec^{\alpha}_{\infty\omega} \mathfrak{B},$$

if the above holds for all $\varphi(x_1 \cdots x_n)$ with $qr(\varphi) \leqq \alpha$.

We shall need the following well-known form of the downward Löwenheim-Skolem-Tarski theorem in sections 4 and 5. We say that L is countable if it has a finite or countable number of relation, function and constant symbols.

LEMMA: *Let $\mathfrak{A} = \langle A, \cdots \rangle$ be an infinite structure for a countable language L, let $A_0 \subseteq A$ and let φ be a sentence with $|\varphi| \leqq |A|$. There is a substructure $\mathfrak{B} = \langle B, \cdots \rangle$ of $\mathfrak{A}$ with $A_0 \subseteq B$ and $|B| =$ $\max\{|\varphi|, |A_0|, \aleph_0\}$ such that for all $\psi(x_1 \cdots x_n) \in \mathrm{sub}(\varphi)$ and all $b_1 \cdots b_n \in B$*

$$\mathfrak{B} \models \psi[b_1 \cdots b_n] \quad \text{iff} \quad \mathfrak{A} \models \psi[b_1 \cdots b_n].$$

Proof: The proof is so similar to the usual proof for L that we shall be content with a sketch. We can assume that $|A_0| \geqq \aleph_0$ and $|A_0| \geqq |\varphi|$ by just enlarging A_0 a bit if necessary. We wellorder A using the axiom of choice. Let $\Phi = \mathrm{sub}(\varphi) \cup \{\neg\psi : \psi \in \mathrm{sub}(\varphi)\}$. Define a sequence $A_0 \subseteq A_1 \subseteq \cdots \subseteq A_n \subseteq \cdots$ such that $a \in A_{n+1}$ iff for some $\psi(x_1 \cdots x_k, y) \in \Phi$ and some $b_1 \cdots b_k \in A_n$

$$\mathfrak{A} \models \exists y \psi(y)[b_1 \cdots b_k],$$

and a is the first element of A such that

$$\mathfrak{A} \models \psi[b_1 \cdots b_k, a].$$

Letting $B = \bigcup_n A_n$, we have the substructure $\mathfrak{B}$ of $\mathfrak{A}$ with universe B, and $|B| = |A_0|$. It is easy to check by induction on formulas that if $\psi \in \Phi$ and $b_1 \cdots b_n \in B$ then:

$$\mathfrak{B} \models \psi[b_1 \cdots b_n] \quad \text{iff} \quad \mathfrak{A} \models \psi[b_1 \cdots b_n]. \dashv$$

3. KARP'S THEOREM

The connecting link between the first two sections is the following very basic result. It was first proved by Karp in [8].

THEOREM 3: *Given structures* $\mathfrak{A}$ *and* $\mathfrak{B}$ *for the language* L, *the following are equivalent:*

(a) $\mathfrak{A} \equiv_{\infty\omega} \mathfrak{B}$
(b) $\mathfrak{A} \cong_p \mathfrak{B}$
(c) *There is a set* I *with* $I: \mathfrak{A} \cong_p \mathfrak{B}$ *such that every* $f \in I$ *has finitely generated domain and range.*

Proof: To prove that (b) implies (a), let $I: \mathfrak{A} \cong_p \mathfrak{B}$. We show by induction on formulas $\varphi(x_1 \cdots x_n)$ of $L_{\infty\omega}$ that if $f \in I$ and $a_1 \cdots a_n \in \text{dom}(f)$ then

$$\mathfrak{A} \models \varphi[a_1 \cdots a_n] \text{ iff } \mathfrak{B} \models \varphi[f(a_1), \cdots, f(a_n)].$$

For φ atomic this is just because f is an isomorphism of its domain and range. The case for propositional connectives is too simple to waste good ink. Suppose that we know the equivalence for $\varphi(x_1 \cdots x_n)$. Let $a_1 \cdots a_{n-1} \in \text{dom}(f)$, $f \in I$. The following are then equivalent:

$\mathfrak{A} \models \exists x_n \varphi[a_1 \cdots a_{n-1}]$.

$\mathfrak{A} \models \varphi[a_1 \cdots a_{n-1}, a]$ for some $a \in A$.

$\mathfrak{A} \models \varphi[a_1 \cdots a_{n-1}, a]$ for some $g \in I$, some $a \in \text{dom}(g)$, $f \subseteq g$.

$\mathfrak{B} \models \varphi[g(a_1) \cdots g(a_{n-1}), g(a)]$ for some $g \in I$,
some $a \in \text{dom}(g)$, $f \subseteq g$.

$\mathfrak{B} \models \varphi[f(a_1) \cdots f(a_{n-1}), b]$ for some $b \in B$.

$\mathfrak{B} \models \exists x_n \varphi[f(a_1) \cdots f(a_{n-1})]$.

The second and fourth equivalences follow from back and forth property; the middle equivalence from the inductive hypothesis on φ. We don't need to check $\forall x_n \varphi$ since this is equivalent to $\neg \exists x_n \neg \varphi$.

Now let us assume $\mathfrak{A} \equiv_{\infty\omega} \mathfrak{B}$ and construct a set I satisfying (c). The above proof of (b) $\Rightarrow$ (a) tells us what I should be. We let $f \in I$ just in case f is an isomorphism of a finitely generated substructure $\mathfrak{A}_0$ of $\mathfrak{A}$ onto a finitely generated substructure $\mathfrak{B}_0$ of $\mathfrak{B}$ such that for all $a_1 \cdots a_n \in A_0$ and all $\varphi(x_1 \cdots x_n)$ of $L_{\infty\omega}$

$$\mathfrak{A} \models \varphi[a_1 \cdots a_n] \quad \text{iff} \quad \mathfrak{B} \models \varphi[f(a_1) \cdots f(a_n)].$$

The assumption that $\mathfrak{A} \equiv_{\infty\omega} \mathfrak{B}$ insures that the substructures of $\mathfrak{A}$ and $\mathfrak{B}$ generated by the empty set are isomorphic and that the isomorphism is in I. To prove that I has the back and forth property without getting bogged down in details, let's assume that our language L has no function symbols and at most a finite number of constant symbols. Then finitely generated just means finite in the above. (If the reader is not inclined to prove the general case for himself he can find it in [1].) So let $f \in I$ and $a \in A$. The case where $b \in B$ is dual. Let $A_0 = \text{dom}(f) = \{a_1, \cdots, a_{n-1}\}$ where $n \geqq 0$. If $a \in A_0$ then we can let $g = f$, so suppose that $a \notin A_0$. We need to find a $b \in B$ such that for all formulas $\varphi(x_1 \cdots x_n)$, if

$$\mathfrak{A} \models \varphi[a_1 \cdots a_{n-1}, a],$$

then

$$\mathfrak{B} \models \varphi[f(a_1) \cdots f(a_{n-1}), b].$$

If there were no such b, then for every $b \in B$ there would be a $\varphi_b(x_1 \cdots x_n)$ with

$$\mathfrak{A} \models \varphi_b[a \cdots a_{n-1}, a] \quad \text{and} \quad \mathfrak{B} \models \neg\varphi_b[f(a_1) \cdots f(a_{n-1}), b].$$

But then letting $\psi(x_1 \cdots x_n)$ equal $\bigwedge\{\varphi_b(x_1 \cdots x_n) : b \in B\}$ we have

$$\mathfrak{A} \models \psi[a_1 \cdots a_{n-1}, a],$$

hence

$$\mathfrak{A} \models \exists x_n \psi(x_n)[a_1 \cdots a_{n-1}]$$

but

$$\mathfrak{B} \models \neg\exists x_n \psi(x_n)[f(a_1) \cdots f(a_{n-1})].$$

This contradicts the definition of I since $f \in I$. Thus, there is a b of the desired type and we can let $g = f \cup \{\langle a, b\rangle\}$ be the extension of f sending a to b. $\dashv$

Applications.

1. The transitivity of the relation $\cong_p$ is less transparent than that of $\equiv_{\infty\omega}$. Also see section 7, where Theorem 3 is used to see that $\cong_p$ is "absolute".

2. If $\mathfrak{A} \equiv_{\infty\omega} \mathfrak{B}$ then $\mathfrak{A}$ and $\mathfrak{B}$ have, up to isomorphism, the same countably generated substructures. For let $I: \mathfrak{A} \cong_p \mathfrak{B}$, $\mathfrak{A}_0 \subseteq \mathfrak{A}$,

where $\mathfrak{A}_0$ is countable. Enumerate $\mathfrak{A}_0$ and construct an embedding of $\mathfrak{A}_0$ into $\mathfrak{B}$ by means of an $f = \bigcup_n f_n$ where $f_n \in I$ and the nth element of $\mathfrak{A}_0$ is in dom (f_n).

3. The cartesian product $\prod_J \mathfrak{A}_j$ of structures is defined in the usual way. For example, if $\mathfrak{A}_j = \langle A_j, R_j \rangle$ with R_j a binary relation on A_j then $\prod_J \mathfrak{A}_j$ is $\langle A, R \rangle$ where $A = \prod_J A_j$ is the cartesian product of the A_j and

$$(\langle a_j \rangle_J, \langle b_j \rangle_J) \in R \quad \text{iff} \quad (a_j, b_j) \in R_j \quad \text{for all } j \in J.$$

If

$$I_j \colon \mathfrak{A}_j \cong_p \mathfrak{B}_j \qquad \text{for all } j \in J,$$

then it is easy to check that

$$I \colon \prod_J \mathfrak{A}_j \cong_p \prod_J \mathfrak{B}_j,$$

where $f \in I$ iff there are $f_j \in I_j$ such that

$$\text{dom}\,(f) = \prod_J \text{dom}\,(f_j) \qquad \text{and} \qquad f(\langle a_j \rangle_J) = \langle f_j(a_j) \rangle_J.$$

Thus, $\mathfrak{A}_j \equiv_{\infty\omega} \mathfrak{B}_j$ for all $j \in J$ implies $\prod_J \mathfrak{A}_j \equiv_{\infty\omega} \prod_J \mathfrak{B}_j$.

4. If L has a constant symbol $\mathbf{0}$ then the direct sum $\sum_J \mathfrak{A}_j$ of structures is defined to be the substructure of $\prod_J \mathfrak{A}_j$ generated by the sequences $\langle a_j \rangle_J$ such that $a_j = 0_j$ for all but a finite number of $j \in J$. (Here 0_j is the interpretation of $\mathbf{0}$ in $\mathfrak{A}_j$.) Using part (c) of Theorem 3 one shows as in application 2 that $\mathfrak{A}_j \equiv_{\infty\omega} \mathfrak{B}_j$ for all $j \in J$ implies $\sum_J \mathfrak{A}_j \equiv_{\infty\omega} \sum_J \mathfrak{B}_j$.

5. Let $\mathscr{C}$ be some category of structures for L which is closed under substructures and let $\mathscr{D}$ be the category of all structures for a language L'. A functor $F \colon \mathscr{C} \to \mathscr{D}$ is *ω-local* if:

(i) $\mathfrak{A} \subseteq \mathfrak{B}$ implies $F(\mathfrak{A}) \subseteq F(\mathfrak{B})$ for $\mathfrak{A}, \mathfrak{B} \in \mathscr{C}$,
(ii) F preserves direct limits.

Suppose that $\mathfrak{A}, \mathfrak{B} \in \mathscr{C}$ and that $I \colon \mathfrak{A} \cong_p \mathfrak{B}$ where by Theorem 3 (c) we have chosen I so that each $f \in I$ has finitely generated domain and range. Define

$$I_F = \{F(f) \colon f \in I\}.$$

This makes sense because F is a functor on $\mathscr{C}$ and $\mathscr{C}$ is closed under substructures. It is not difficult to see that if F is ω-local, then I_F has the back and forth property so that $F(\mathfrak{A}) \cong_p F(\mathfrak{B})$. Applying Theorem 3 we see that for $\mathfrak{A}$, $\mathfrak{B} \in \mathscr{C}$, if $\mathfrak{A} \cong_{\infty\omega} \mathfrak{B}$ then $F(\mathfrak{A}) \equiv_{\infty\omega} F(\mathfrak{B})$. This is a simple special case of a result in Feferman [16].

THEOREM 4: *For any two structures $\mathfrak{A}$ and $\mathfrak{B}$ for the language L with $\mathfrak{A} \subseteq \mathfrak{B}$, the following are equivalent:*

(a) $\mathfrak{A} \prec_{\infty\omega} \mathfrak{B}$
(b) *There is a set I such that $I\colon \mathfrak{A} \cong_p \mathfrak{B}$ and for every $a_1, \cdots, a_n \in A$ there is an $f \in I$ with $f(a_i) = a_i$ for $i = 1, \cdots, n$.*

This theorem follows from the proof of Theorem 3, as the reader can observe by rereading the earlier proof. Example 3 of section 1 provides an interesting application of this theorem since it yields $G \prec_{\infty\omega} H$.

Application.

6. Let L have a constant symbol $\mathbf{0}$. For any index set J we write $\oplus_J\mathfrak{A}$ for $\sum_J \mathfrak{A}_j$ where $\mathfrak{A}_j = \mathfrak{A}$ for all $j \in J$. Using Theorem 4 we can show that if J and J' are infinite, $J \subseteq J'$ then

$$\oplus_J\mathfrak{A} \prec_{\infty\omega} \oplus_{J'}\mathfrak{A}.$$

To see this, let $\mathfrak{B} = \oplus_J \mathfrak{A}$ and $\mathfrak{C} = \oplus_{J'} \mathfrak{A}$. For $J_0 \subseteq J$ let $\mathfrak{B}(J_0)$ be the substructure of $\mathfrak{B}$ generated by those sequences $\langle a_j \rangle$ such that $a_j = 0$ for $j \notin J_0$, and define $\mathfrak{C}(J_1)$ similarly for $J_1 \subseteq J'$. Let $f \in I$ iff there are finite sets $J_0 \subseteq J$, $J_1 \subseteq J'$ such that

$$f\colon \mathfrak{B}[J_0] \cong \mathfrak{C}(J_1).$$

Then $I\colon \mathfrak{B} \cong_p \mathfrak{C}$ and for any finite $J_0 \subseteq J$, the restriction of the identity map to $\mathfrak{B}[J_0]$ is in I.

The following, also due to Karp, is a refined version of Theorem 3. It gives a back and forth characterization of $\equiv^{\alpha}_{\infty\omega}$ for any ordinal α, which one might write naturally as $\cong^{\alpha}_p$. It suggests unexplored regions between algebra and logic, one of which is treated briefly in section 6.

THEOREM 5: *For any two structures* $\mathfrak{A}$ *and* $\mathfrak{B}$ *for the language* L *and any ordinal* α *the following are equivalent:*

(a) $\mathfrak{A} \equiv^{\alpha}_{\infty\omega} \mathfrak{B}$
(b) *There is a sequence*

$$I_0 \supseteq I_1 \supseteq \cdots \supseteq I_\beta \supseteq \cdots \supseteq I_\alpha,$$

where, for each $\beta \leqq \alpha$, I_β *is a nonempty set of partial isomorphisms between (finitely generated) substructures of* $\mathfrak{A}$ *and* $\mathfrak{B}$, *and such that if* $\beta + 1 \leqq \alpha$ *and* $f \in I_{\beta+1}$, *then for any* $a \in A$ $(b \in B)$ *there is a* $g \in I_\beta$ *with* $f \subseteq g$ *and* $a \in \mathrm{dom}\,(g)$ *(resp.* $b \in \mathrm{rng}(g)$*).*

Proof: The proof is a rather routine modification of the proof of Theorem 3. To prove (b) implies (a) show by induction on β that $f \in I_\beta$, $a_0 \cdots a_{n-1} \in \mathrm{dom}\,(f)$ and $qr(\varphi(v_0 \cdots v_{n-1})) \leqq \beta$ implies

$$\mathfrak{A} \models \varphi[a_0 \cdots a_{n-1}] \quad \text{iff} \quad \mathfrak{B} \models \varphi[f(a_0) \cdots f(a_{n-1})].$$

This fact dictates what I_β should be in proving (a) implies (b). ⊣

All of applications 3–5 above have analogues for $\equiv^{\alpha}_{\infty\omega}$ which we leave to the reader. There is also a result similar to Theorem 4 for $\prec^{\alpha}_{\infty\omega}$, also left to the reader.

4. SCOTT'S THEOREM

If we combine Theorems 2 and 3 we see that if $\mathfrak{A}$ and $\mathfrak{B}$ are countable structures for our language L then $\mathfrak{A} \equiv_{\infty\omega} \mathfrak{B}$ iff $\mathfrak{A}$ and $\mathfrak{B}$ are isomorphic. In this section we strengthen this result considerably in the case where L is countable, that is, has at most countably many relation, function and constant symbols. (In this case countable and countably generated coincide.) The result mentioned is Corollary 1 following Theorem 7 and is commonly referred to as Scott's Theorem.

THEOREM 6: *Let* $\mathfrak{A}$ *be a structure for the language* L. *For any* $a_1 \cdots a_n \in A$ *and any ordinal* α *there is a formula* $\varphi(v_1 \cdots v_n)$ *of* $L_{\infty\omega}$ *with* $qr(\varphi) = \alpha$ *such that for any structure* $\mathfrak{B}$ *and any* $b_1 \cdots b_n \in B$, *the following are equivalent:*

1. $(\mathfrak{A}, a_1 \cdots a_n) \equiv^{\alpha}_{\infty\omega} (\mathfrak{B}, b_1 \cdots b_n)$.
2. $\mathfrak{B} \models \varphi[b_1 \cdots b_n]$.

Proof: For any sequence $s = \langle a_1 \cdots a_n \rangle$ $(n \geqq 0)$ and any ordinal α we define a formula $\varphi^{\alpha}_{s}(v_1 \cdots v_n)$ by recursion on α.

$\varphi^{0}_{s}(v_1 \cdots v_n)$ is the conjunction of all atomic or negated atomic formulas θ with free variables among $v_1 \cdots v_n$ such that

$$\mathfrak{A} \models \theta\,[a_1 \cdots a_n].$$

If $s = \langle a_1 \cdots a_n \rangle$ then we use $s \frown a$ for the sequence $\langle a_1 \cdots a_n, a \rangle$. For any α, $\varphi^{\alpha+1}_{s}(v_1 \cdots v_n)$ is the conjunction of

(i) $\varphi^{\alpha}_{s}(v_1 \cdots v_n)$,
(ii) $\bigwedge_{a \in A} \exists v_{n+1} \varphi^{\alpha}_{s \frown a}(v_1 \cdots v_{n+1})$, and
(iii) $\forall v_{n+1} \bigvee_{a \in A} \varphi^{\alpha}_{s \frown a}(v_1 \cdots v_{n+1})$.

For limit ordinals, λ, φ^{λ}_{s} is $\bigwedge_{\alpha < \lambda} \varphi^{\alpha}_{s}$. Note that $qr(\varphi^{\alpha}_{s}) = \alpha$ is provable by a simple induction on α. Also note that if $s = \langle a_1 \cdots a_n \rangle$ then $\mathfrak{A} \models \varphi^{\alpha}_{s}[a_1 \cdots a_n]$. This implies that if

$$(\mathfrak{A}, a_1 \cdots a_n) \equiv^{\alpha}_{\infty\omega} (\mathfrak{B}, b_1 \cdots b_n),$$

then

$$\mathfrak{B} \models \varphi^{\alpha}_{s}[b_1 \cdots b_n].$$

Suppose, to prove the converse, that $\mathfrak{B} \models \varphi^{\alpha}_{s_0}[b_1 \cdots b_n]$ where $s_0 = \langle a_1 \cdots a_n \rangle$. For $\beta \leqq \alpha$ let $f \in I_\beta$ iff f is a mapping of a finitely generated substructure of $(\mathfrak{A}, a_1 \cdots a_n)$, generated by say $c_1 \cdots c_k$, onto a finitely generated substructure of $(\mathfrak{B}, b_1 \cdots b_n)$ such that

$$\mathfrak{B} \models \varphi^{\beta}_{s}[f(c_1) \cdots f(c_k)],$$

where $s = \langle c_1 \cdots c_k \rangle$. Since $\mathfrak{B} \models \varphi^{\alpha}_{s_0}[b_1 \cdots b_n]$, the trivial map is in I_α. Since

$$\varphi^{\beta}_{s}(v_1 \cdots v_k) \rightarrow \varphi^{\gamma}_{s}(v_1 \cdots v_k)$$

is logically valid whenever $\gamma \leqq \beta$, $I_\beta \subseteq I_\gamma$ for $\gamma \leqq \beta$. Suppose that

$$f \in I_{\beta+1}, \qquad \beta + 1 \leqq \alpha \quad \text{and} \quad a \in A.$$

Thus,

$$\mathfrak{B} \models \varphi^{\beta+1}_{s}[f(c_1) \cdots f(c_k)],$$

and in particular, by (ii),

$$\mathfrak{B} \models \exists v_{k+1} \varphi^{\beta}_{s\frown a}(v_{k+1})[f(c_1) \cdots f(c_k)].$$

Let d be such a v_{k+1}. Let $g(a) = d$ and extend to an isomorphism of the structures generated by $c_1 \cdots c_k$, a and $f(c_1) \cdots f(c_k)$, d in the natural way. Then $g \in I_\beta$ and $a \in \mathrm{dom}\,(g)$. If $f \in I_{\beta+1}$ and $b \in B$ we can get a $g \in I_\beta$ with $b \in \mathrm{rng}\,(g)$ by using (iii). Thus the sequence of I_β has the properties required by Theorem 5 to show that

$$(\mathfrak{A}, a_1 \cdots a_n) \equiv^{\alpha}_{\infty\omega} (\mathfrak{B}, b_1 \cdots b_n). \dashv$$

THEOREM 7: *Let $\mathfrak{A}$ be a structure for L. For any $a_1 \cdots a_n \in A$ there is a formula $\varphi(v_1 \cdots v_n)$ of $L_{\infty\omega}$ with $|\varphi| \leqq \max\{|A|, |L|, \aleph_0\}$ such that for any $\mathfrak{B}$ and any $b_1 \cdots b_n \in B$, the following are equivalent:*

(1) $(\mathfrak{A}, a_1 \cdots a_n) \equiv_{\infty\omega} (\mathfrak{B}, b_1 \cdots b_n)$.
(2) $\mathfrak{B} \models \varphi[b_1 \cdots b_n]$.

(For $n = 0$, this sentence is referred to as the *Scott sentence* of $\mathfrak{A}$ and is denoted by $\sigma(\mathfrak{A})$. Thus, the Scott sentence of $\mathfrak{A}$ determines $\mathfrak{A}$ up to $\infty\omega$-equivalence. In general, we denote the formula $\varphi(x_1 \cdots x_n)$ of Theorem 7 by $\sigma^{\infty}_{\langle a_1 \cdots a_n \rangle}(x_1 \cdots x_n)$.)

Proof: Let κ be the least ordinal with $|\kappa| > \max\{|A|, \aleph_0\}$. Let $\varphi^{\alpha}_{s}(v_1 \cdots v_n)$ be the formula constructed in the proof of Theorem 6. For a fixed sequence s, the sets

$$X^{\alpha}_{s} = \{(c_1 \cdots c_k) \colon \mathfrak{A} \models \varphi^{\alpha}_{s}[c_1 \cdots c_k]\},$$

form a decreasing sequence of subsets of A^k so there is at least ordinal $\beta(s) < \kappa$ such that $X^{\beta(s)}_{s} = X^{\alpha}_{s}$ for all $\alpha \geqq \beta(s)$. That is,

$$\mathfrak{A} \models \forall v_1 \cdots \forall v_k [\varphi^{\beta(s)}_{s} \leftrightarrow \varphi^{\alpha}_{s}],$$

for all $\alpha \geqq \beta(s)$. The ordinal

$$\gamma = \sup\{\beta(s) \colon s \text{ a finite sequence extending } \langle a_1 \cdots a_n \rangle\},$$

is also less than κ, and $a_1 \cdots a_n$ satisfies the conjunction $\psi(v_1 \cdots v_n)$ (over all sequences $s = \langle a_1 \cdots a_n, c_1 \cdots c_k \rangle$ from A extending $\langle a_1 \cdots a_n \rangle$) of:

$$\forall v_{n+1} \cdots \forall v_{n+k} [\varphi^{\gamma}_{s} \leftrightarrow \varphi^{\gamma+1}_{s}].$$

Let $\varphi(v_1 \cdots v_n)$ be

$$\varphi^{\gamma}_{\langle a_1 \cdots a_n \rangle}(v_1 \cdots v_n) \wedge \psi(v_1 \cdots v_n).$$

The cardinality condition on φ is easily checked. We know that $\mathfrak{A} \models \varphi[a_1 \cdots a_n]$ so if

$$(\mathfrak{A}, a_1 \cdots a_n) \equiv_{\infty\omega} (\mathfrak{B}, b_1 \cdots b_n),$$

then

$$\mathfrak{B} \models \varphi[b_1 \cdots b_n].$$

For the converse, suppose that $\mathfrak{B} \models \varphi[b_1 \cdots b_n]$ and define I_β for $\beta < \kappa$ as in the proof of Theorem 6. Since

$$\mathfrak{B} \models \forall v_{n+1} \cdots \forall v_{n+k}[\varphi^{\gamma}_s \leftrightarrow \varphi^{\gamma+1}_s][b_1 \cdots b_n],$$

for each s, we see that $I_\gamma = I_{\gamma+1}$. Thus

$$I_\gamma\colon (\mathfrak{A}, a_1 \cdots a_n) \cong_p (\mathfrak{B}, b_1 \cdots b_n),$$

and the result follows from Theorem 3. ⊣

Technical Remark: The ordinal γ is an important invariant of $\langle a_1 \cdots a_n \rangle$ and $\mathfrak{A}$. It is $\leqq$ the least non-hyperprojective (over $\mathfrak{A}$) ordinal in the sense of Moschovakis [10] since the set

$$Y = \{\langle a_1 \cdots a_n \rangle, \langle b_1 \cdots b_n \rangle)\colon (\mathfrak{A}, a_1 \cdots a_n) \not\equiv_{\infty\omega} (\mathfrak{A}, b_1 \cdots b_n)\},$$

is given by a first order positive inductive definition over $\mathfrak{A}$. It also follows that the relation $(s, s') \in Y$ is semi-hyperprojective.

DEFINITION: The set of formulas φ with $|\varphi| \leqq \aleph_0$ is denoted by $L_{\omega_1\omega}$.

COROLLARY 1: (Scott's Theorem). *Given a countable structure $\mathfrak{A}$ for a countable language L, there is a sentence φ of $L_{\omega_1\omega}$ such that for any countable structure $\mathfrak{B}$, $\mathfrak{B} \models \varphi$ iff $\mathfrak{A} \cong \mathfrak{B}$.*

Proof: Let $\varphi = \sigma(\mathfrak{A})$ be the Scott sentence of $\mathfrak{A}$ and apply Theorems 2 and 3. ⊣

We can now see that, in a sense, the back and forth method is the only way to show that all countable infinite models of a countable theory T (of $L_{\omega_1\omega}$) are isomorphic.

COROLLARY 2: *A countable theory T of $L_{\omega_1\omega}$ is $\aleph_0$-categorical iff all infinite models of T are partially isomorphic.*

Proof: In view of earlier results, it suffices to show that if $\mathfrak{A} \models T$, $\mathfrak{B} \models T$ and $\mathfrak{A}$ is countable, then $\mathfrak{B} \models \varphi$, where φ is the Scott sentence of $\mathfrak{A}$, under the assumption that T is $\aleph_0$-categorical. Let $\psi = \varphi \wedge \bigwedge T$, and apply the downward Löwenheim-Skolem lemma of section 2 to the model $\mathfrak{B}$ and sentence ψ, to get a countable submodel $\mathfrak{B}_0$ of $\mathfrak{B}$ such that $\mathfrak{B}_0 \models T$ and

$$\mathfrak{B}_0 \models \varphi \quad \text{iff} \quad \mathfrak{B} \models \varphi.$$

But $\mathfrak{B}_0 \cong \mathfrak{A}$ so $\mathfrak{B}_0 \models \varphi$ and hence $\mathfrak{B} \models \varphi$. ⊣

If one takes the relation $\cong_p$ seriously (another reason for doing so is given in section 7) then a number of companion notions arise naturally. For example, f is a *semi-isomorphism* of $\mathfrak{A}$ and $\mathfrak{B}$ if $f \in I$ for some $I: \mathfrak{A} \cong_p \mathfrak{B}$; equivalently, if f is a partial isomorphism preserving satisfaction of $L_{\infty\omega}$ formulas. Notice that the inverse of a semi-isomorphism is a semi-isomorphism as is the composition of two semi-isomorphisms. If $\mathfrak{A}$ and $\mathfrak{B}$ are countable then every semi-isomorphism can be extended to an isomorphism $f: \mathfrak{A} \cong \mathfrak{B}$, by Theorem 2.

If f is a semi-isomorphism of $\mathfrak{A}$ and $\mathfrak{A}$, then f is called a semi-automorphism. An n-ary relation R on $\mathfrak{A}$ is *strongly invariant* iff for every semi-automorphism f of $\mathfrak{A}$, and all $a_1 \cdots a_n \in \text{dom}(f)$:

$$R(a_1 \cdots a_n) \quad \text{iff} \quad R(f(a_1), \cdots, f(a_n)).$$

Thus, for countably generated structures, R is strongly invariant iff R is invariant under all automorphisms. The following is a straightforward generalization of a result of Scott.

COROLLARY 3: *A relation R on $\mathfrak{A}$ is strongly invariant iff there is a formula $\varphi(x_1 \cdots x_n)$ of $L_{\infty\omega}$ such that for all $a_1 \cdots a_n \in A$:*

$$R(a_1 \cdots a_n) \quad \text{iff} \quad \mathfrak{A} \models \varphi[a_1 \cdots a_n].$$

(If $\mathfrak{A}$ and L are countable then $\varphi \in L_{\omega_1\omega}$.)

Proof: Any definable relation is strongly invariant by definition. If R is strongly invariant then let $\varphi(x_1 \cdots x_n)$ be

$$\bigvee\{\sigma^{\infty}_{\langle a_1 \cdots a_n\rangle}(x_1 \cdots x_n) : R(a_1 \cdots a_n)\}.$$

If $\mathfrak{A} \models \varphi[b_1 \cdots b_n]$ then $\mathfrak{A} \models \sigma^{\infty}_{\langle a_1 \cdots a_n \rangle}[b_1 \cdots b_n]$ for some $\langle a_1 \cdots a_n \rangle$ so that $(\mathfrak{A}, a_1 \cdots a_n) \equiv_{\infty\omega} (\mathfrak{A}, b_1 \cdots b_n)$ and hence there is a semi-automorphism f with $f(a_i) = b_i$ (all $i = 1, \cdots n$). Since R is strongly invariant, $R(b_1 \cdots b_n)$. ⊣

A structure $\mathfrak{A}$ is (strongly) rigid if there is no nontrivial (semi-) automorphism of $\mathfrak{A}$. (For countable structures the two notions coincide.) This is equivalent to saying that $R = \{a\}$ is (strongly) invariant for each $a \in A$. Thus, by Corollary 3, $\mathfrak{A}$ is strongly rigid just in case each $a \in A$ is definable by a formula $\varphi(x)$ of $L_{\infty\omega}$ (or $L_{\omega_1\omega}$ if $\mathfrak{A}$ and L are countable). This is of special interest in view of the following result of Kueker [9].

THEOREM 8: *Let $\mathfrak{A}$ be a countable structure such that for any finite number $a_1 \cdots a_n$ of elements of A there is a nontrivial automorphism f of $\mathfrak{A}$ with $f(a_i) = a_i$ for all $i = 1, \cdots, n$. Then $\mathfrak{A}$ has $2^{\aleph_0}$ automorphisms.*

Proof: The hypothesis of the theorem can be restated as: for any $a_1 \cdots a_n \in A$, $(\mathfrak{A}, a_1 \cdots a_n)$ is not rigid. For ease in notation let us write $(a_1 \cdots a_n) \equiv (b_1 \cdots b_n)$ for

$$(\mathfrak{A}, a_1 \cdots a_n) \equiv_{\infty\omega} (\mathfrak{A}, b_1 \cdots b_n).$$

We first show that if $(\mathfrak{A}, a_1 \cdots a_n)$ is not rigid and $(a_1 \cdots a_n) \equiv (b_1 \cdots b_n)$ then there are c, d and d' such that $d \neq d'$ and

$$(a_1 \cdots a_n, c) \equiv (b_1 \cdots b_n, d) \equiv (b_1 \cdots b_n, d').$$

For, since $(\mathfrak{A}, a_1 \cdots a_n)$ is not rigid there is an automorphism g of $\mathfrak{A}$ leaving each a_i fixed but with $g(c) \neq c$ for some c. Now, using the back and forth property there are d and d' such that

$$(a_1 \cdots a_n, c, g(c)) \equiv (b_1 \cdots b_n, d, d'),$$

and hence $d \neq d'$ and

$$(a_1 \cdots a_n, c) \equiv (b_1 \cdots b_n, d),$$
$$(a_1 \cdots a_n, g(c)) \equiv (b_1 \cdots b_n, d').$$

But since g is an automorphism,

$$(a_1 \cdots a_n, c) \equiv (a_1 \cdots a_n, g(c)),$$

so the desired result follows by the transitivity of $\equiv$.

To finish the proof we assume that $(\mathfrak{A}, a_1 \cdots a_n)$ is not rigid for all $a_1 \cdots a_n \in A$ and use the above fact repeatedly to construct $2^{\aleph_0}$ automorphisms of $\mathfrak{A}$. We do this by defining a distinct automorphism f_s for each function $s \in 2^\omega$. For $s \in 2^\omega$ let

$$s_n = s \upharpoonright \{0, 1, \cdots n - 1\}.$$

We define f_s to be $\bigcup_n f(s_n)$ where the $f(s_n)$ are semi-automorphisms (with finitely generated domain and range), defined for all s simultaneously by induction on n. Before defining them let I be the set of all such semi-automorphisms and note that $(a_1 \cdots a_n) \equiv (b_1 \cdots b_n)$ iff there is an $f \in I$ such that $f(a_i) = b_i$ $(i = 1 \cdots n)$.

Let $A = \{a_0, a_1, \cdots\}$ be an enumeration of A, and let $f(s_0) = \varnothing$ be the empty function. Given $f(s_n)$ first extend it to a semi-automorphism $g \in I$ with $a_n \in \text{dom}\,(g) \cap \text{rng}\,(g)$ by using the back and forth property twice. (This insures that $\text{dom}\,(f_s) = \text{rng}\,(f_s) = A$ when we are finished.) Now, given g with domain generated by, say, $c_1 \cdots c_k$ we have

$$(c_1 \cdots c_k) \equiv (g(c_1) \cdots g(c_k)).$$

Using the above remark, we find a c_{k+1}, d and d' so that $d \neq d'$,

$$(c_1 \cdots c_k, c_{k+1}) \equiv (g(c_1) \cdots g(c_k), d),$$

and

$$(c_1 \cdots c_k, c_{k+1}) \equiv (g(c_1) \cdots g(c_k), d').$$

We may thus extend g to two functions g_0 and g_1 in I so that $g_0(c_{k+1}) = d$ and $g_1(c_{k+1}) = d'$. Note that $f(s_n) \subseteq g_0$ and $f(s_n) \subseteq g_1$. We define $f(s_{n+1})$ to be g_0 if $s(n) = 0$ and to be g_1 if $s(n) = 1$. Thus distinct sequences give rise to distinct automorphisms. ⊣

COROLLARY 1: *Let $\mathfrak{A}$ be a countable structure with $< 2^{\aleph_0}$ automorphisms. Then $\mathfrak{A}$ has $\leqq \aleph_0$ automorphisms.*

Proof: Every automorphism of $\mathfrak{A}$ will be determined by what it does to the sequence $a_1 \cdots a_n$ that makes $(\mathfrak{A}, a_1 \cdots a_n)$ rigid. ⊣

COROLLARY 2: *Let $\mathfrak{A}$ be a countable structure for L. $\mathfrak{A}$ has $< 2^{\aleph_0}$ automorphisms iff there are $a_1 \cdots a_n$ such that every element b of $\mathfrak{A}$ is definable in $(\mathfrak{A}, a_1 \cdots a_n)$ by a formula of $\varphi(x)$ of $L_{\infty\omega}$, or $L_{\omega_1\omega}$ if L is countable.*

This is the form in which Theorem 8 is stated in Kueker [9].

COROLLARY 3: *Let* $\mathfrak{A} \prec_{\infty\omega} \mathfrak{B}$ *where* $\mathfrak{A} \neq \mathfrak{B}$. *If* $\mathfrak{A}$ *is countable then* $\mathfrak{A}$ *has* $2^{\aleph_0}$ *automorphisms. Hence, if* $\mathfrak{A} \equiv_{\infty\omega} \mathfrak{B}$ *where* $\mathfrak{A}$ *is countable and* $\mathfrak{B}$ *is uncountable then* $\mathfrak{A}$ *has* $2^{\aleph_0}$ *automorphisms.*

Proof: The first statement is almost immediate from Corollary 2. The second follows from the fact that $\mathfrak{A}$ must be isomorphic to some $\mathfrak{B}_0 \prec_{\infty\omega} \mathfrak{B}$. $\dashv$

Technical Remark: Using the remark following Theorem 7, many of these results can be improved by getting bounds on the formula of $L_{\infty\omega}$ asserted to exist in the above. For example, M. Nadel has shown that $\mathfrak{A} \prec_{\infty\omega} \mathfrak{B}$ can be replaced by $\mathfrak{A} \prec^{\kappa}_{\infty\omega} \mathfrak{B}$ in Corollary 3, where κ is the least ordinal not hyperprojective over $\mathfrak{A}$. This is better since $\mathfrak{A} \prec_{\infty\omega} \mathfrak{B}$ implies $\mathfrak{A} \cong \mathfrak{B}$ (for countable $\mathfrak{A}$ and $\mathfrak{B}$) but $\mathfrak{A} \prec^{\kappa}_{\infty\omega} \mathfrak{B}$ is in general weaker.

5. $\aleph_1$-FREE ABELIAN GROUPS

The theory of Abelian groups is full of hidden (and not so hidden) back and forth arguments. We give two examples in this and the next section.

An Abelian group G is $\aleph_1$-*free* if every countable subgroup of G is free. Since every subgroup of a free group is free, every free group is $\aleph_1$-free. To make the statement of the following simpler, we modify the usual definition by declaring that no free group on a finite number of generators is $\aleph_1$-free.

The following result is due to Kueker. His original back and forth argument lies buried beneath our machinery.

THEOREM 9: *Let* G *be the free group on* $\aleph_0$*-generators. Then for any* H, $G \cong_p H$ *iff* H *is* $\aleph_1$*-free.*

Proof: Suppose that $G \cong_p H$, and that H_0 is a countable subgroup of H. We need to prove that H_0 is free, or, what is the same, isomorphic to a subgroup of G. This follows from Application 2 of Theorem 3.

To prove the converse, suppose that H is $\aleph_1$-free. To show that

$G \cong_p H$ it suffices to show that H is a model of the Scott sentence $\sigma(G)$. Let A be an infinite set of independent elements of H. Using the Löwenheim-Skolem lemma, we get a countable $H_1 \subseteq H$, $A \subseteq H_1$, such that $H \models \sigma(G)$ iff $H_1 \models \sigma(G)$. But $H_1 \cong G$ so $H_1 \models \sigma(G)$. ⊣

Application: Let G and H be as in Example 3 of section 1. Since $G \prec_{\infty\omega} H$, H is $\aleph_1$-free. (The usual proof that H is $\aleph_1$-free contains the above proof that $G \prec_{\infty\omega} H$ as a hidden subproof.) It is not difficult to see that H is not free. Thus, we see that the notion of free Abelian group is not expressible in $L_{\infty\omega}$. (This was also noticed by H. J. Keisler, though his proof was different from Kueker's.) On the other hand, the notion of $\aleph_1$-free group is expressible in $L_{\omega_1\omega}$, say by the sentence $\sigma(G)$.

The reader might be interested in seeing a more group theoretic definition of $\aleph_1$-free which is easily expressible in $L_{\omega_1\omega}$.

COROLLARY: *An Abelian group H is $\aleph_1$-free just in case it satisfies the following conditions, each of which is expressible in $L_{\omega_1\omega}$:*

(1) *H is torsion free,*
(2) *H is not free on a finite number of generators,*
(3) *For each $x_1 \cdots x_n$, the pure subgroup of H generated by $x_1 \cdots x_n$ is free on a finite number of generators.*

Proof: Let G be a free Abelian group on an infinite number of generators. Let $f \in I$ iff f is an isomorphism of a finitely generated pure subgroup of G onto a finitely generated pure subgroup of H. The conditions (1)–(3) insure that $I: G \cong_p H$, so H is $\aleph_1$-free by Theorem 9. ⊣

The above Corollary is really just a restatement of Pontryagin's Criterion (cf. Fuch's [5], p. 51).

6. TORSION GROUPS

We turn from torsion free groups to Abelian *torsion* groups. The fundamental theorem in the theory of torsion groups is Ulm's

Theorem. The net result of Ulm's Theorem is a complete set of invariants for countable (the magic word) torsion groups.

We first write any torsion group G uniquely as a direct sum

$$\sum_p G_p,$$

over all primes p, where G_p is a p-group, i.e., a model of

$$\forall x \bigvee_n [p^n \cdot x = \mathbf{0}].$$

Given a p-group G we write

$$G = G_r \oplus G_d,$$

uniquely, where G_d is divisible (i.e., a model of $\forall x \exists y(x = p \cdot y)$) and G_r is reduced (i.e., G_r has no divisible subgroups except $\{0\}$). Now G_d is just a direct sum of a certain number of copies of $Z(p^\infty)$, the group of all p^nth complex roots of unity, as n varies over the positive integers, so to characterize torsion groups, it suffices to characterize reduced p-groups, for the various primes p.

Let p be a fixed prime. We now use "group" to mean Abelian p-group. We define

$$pG = \{p \cdot x : x \in G\} = \{y \in G : G \models \exists x[y = px]\}.$$

Then for any ordinal α, we define $p^\alpha G$ as follows:

$$p^0 G = G,$$

$$p^{\alpha+1}G = p(p^\alpha G),$$

$$p^\lambda G = \bigcap_{\beta<\lambda} p^\beta G,$$

if λ is a limit ordinal. By a cardinality argument, there is an ordinal τ, of cardinality at most that of G, such that $p^\tau G = p^{\tau+1}G$ and then $p^\tau G = \bigcap_\alpha p^\alpha G$. In general, this $p^\tau G$ is the divisible part G_d of G. If G is reduced then $p^\tau G = \{0\}$. This least τ is called the *length* of G, $\tau = l(G)$.

We wish to associate with each $\alpha < l(G)$ a cardinal number $U_G(\alpha)$, the αth Ulm invariant of G. To do this, define $G[p]$ to be the set of elements x of G with $px = 0$. (We write $p^\alpha G[p]$ instead of $(p^\alpha G)[p]$.) If $G[p] = G$, that is, if every $x \in G$ has order p, then we can regard G as a vector space over Z_p, the field with p elements, and G will have a certain dimension. In particular,

$$p^{\alpha}G[p]/p^{\alpha+1}G[p],$$

has a dimension. It is this dimension which we take for $U_G(\alpha)$. For $\alpha \geqq l(G)$, $U_G(\alpha) = 0$. We define $\hat{U}_G(\alpha)$ to be $U_G(\alpha)$ if $U_G(\alpha)$ is finite, to be the symbol ∞ if $U_G(\alpha)$ is infinite. Ulm's Theorem states that if G and H are countable reduced p-groups, then $G \cong H$ iff $U_G(\alpha) = U_H(\alpha)$ for all α. However, if we look at the proof of this result in Kaplansky [7], we see that it really shows the hard half of the following:

THEOREM 10: *For any two reduced p-groups G and H, $G \cong_p H$ iff $\hat{U}_G(\alpha) = \hat{U}_H(\alpha)$ for all α.*

Proof of easy half: If $G \cong_p H$ then $G \equiv_{\infty\omega} H$ by Theorem 3. But the statement "$p^{\alpha}G[p]$ has n elements independent mod $p^{\alpha+1}G[p]$" can easily be expressed by a sentence $\theta_{\alpha,n}$ of $L_{\infty\omega}$. Then $\hat{U}_G(\alpha) = n$ iff $G \models \theta_{\alpha,n} \wedge \neg\theta_{\alpha,n+1}$, and $\hat{U}_G(\alpha) = \infty$ iff $G \models \bigwedge_n \theta_{\alpha,n}$. Thus $G \equiv_{\infty\omega} H$ implies $\hat{U}_G(\alpha) = \hat{U}_H(\alpha)$ for all α. ⊣

If one is careful, one can write the above $\theta_{\alpha,n}$ so that its quantifier rank is $\delta + n + 1$ where $\alpha = \omega \cdot \delta + m$ for some $m < \omega$. Then $\hat{U}_G(\alpha) = n$ will be expressed by a sentence of quantifier ranks $\delta + n + 2$, and $\hat{U}_G(\alpha) = \infty$ will be expressed by a sentence of quantifier rank $\delta + \omega$. In particular, if $\beta = \omega\lambda$, where λ is a limit ordinal, then for each $\alpha < \beta$, $\hat{U}_G(\alpha)$ can be defined by a sentence of quantifier rank $<\lambda$. Thus, if

$$G \equiv^{\lambda}_{\infty\omega} H,$$

then

$$\hat{U}_G(\alpha) = \hat{U}_H(\alpha) \quad \text{for all } \alpha < \omega \cdot \lambda.$$

It is a rather remarkable fact that the converse is also true. This gives a precise form to the intuition that the further out $U_G(\alpha) = U_H(\alpha)$, the more indistinguishable G and H are. This is a simplified version of a result in Barwise-Eklof [1].

THEOREM 11: *Let G and H be reduced p-groups and let δ be an ordinal. If $\hat{U}_G(\alpha) = \hat{U}_H(\alpha)$ for all $\alpha < \omega \cdot \delta$ then $G \equiv^{\delta}_{\infty\omega} H$. If δ is a limit ordinal, then the converse holds.*

Proof: We will need to assume that the reader is familiar with section 11 of Kaplansky [7]. Assume that $\hat{U}_G(\alpha) = \hat{U}_H(\alpha)$ for all $\alpha < \omega \cdot \delta$. If $l(G) < \omega \cdot \delta$ then $\hat{U}_H(\alpha) = 0$ for all α with $l(G) \leqq \alpha < \omega \cdot \delta$ so $l(H) = l(G)$ and hence $\hat{U}_G(\alpha) = \hat{U}_H(\alpha)$ for all α. Thus $G \equiv_{\infty\omega} H$ by Theorem 10. The hard part of the proof is when one, and hence both, of $l(G)$, $l(H)$ are $\geqq \omega \cdot \delta$, which we now assume. In order to apply Theorem 5 we define a sequence

$$I_0 \supseteq \cdots \supseteq I_\beta \supseteq \cdots \supseteq I_\delta \qquad (\beta \leqq \delta)$$

of sets of isomorphisms as follows: $f \in I_\beta$ iff f is an isomorphism of a finite subgroup S of G onto a finite subgroup T of H such that for all $x \in S$, if

$$h_G(x) < \omega\beta \qquad \text{then} \qquad h_G(x) = h_H(f(x)),$$

and if

$$h_G(x) \geqq \omega\beta \qquad \text{then} \qquad h_H(f(x)) \geqq \omega\beta.$$

Here $h_G(x)$ is the height of x in G, i.e., the least α such that $x \in p^\alpha G - p^{\alpha+1}G$. For $x = 0$, $h_G(0) = \infty$ and we define $\alpha < \infty$ for all α. Thus, I_β is the set of those isomorphisms which preserve heights less than $\omega\beta$. We need to show that if $f \in I_{\beta+1}$, $a \in G$ then there is a $g \in I_\beta$ such that $f \subseteq g$ and $a \in \text{dom}\,(g)$. The dual condition follows by symmetry. Let $S = \text{dom}\,(f)$, $T = \text{rng}(f)$ and assume $a \notin S$. Let r be the least integer $\geqq 0$ such that $p^{r+1} a \in S$. Such an r must exist since G is a p-group. The proof breaks into cases.

(a) $h_G(a) \geqq \omega\beta$. Since f preserves heights less than

$$\omega \cdot (\beta + 1) = \omega \cdot \beta + \omega,$$

and since

$$h_G(p^{r+1}a) \geqq h_G(a) + r + 1 \geqq \omega\beta + r + 1,$$

we have

$$h_H(f(p^{r+1}a)) \geqq \omega\beta + r + 1.$$

Thus, there is a $b_0 \in p^{\omega\beta}H$ such that

$$p^{r+1}b_0 = f(p^{r+1}a).$$

If $p^r b_0 \notin T$ let $b = b_0$. If $p^r b_0 \in T$ then one has to find a $b_1 \in p^{\omega\beta}H$ with $p^{r+1}b_1 = 0$ and $p^r b_1 \notin T$. This is not difficult since T is finite and $l(H) \geqq \omega\beta + \omega$. Then let $b = b_0 + b_1$. In either case,

$$h_H(b) \geqq \omega\beta, \qquad p^{r+1}b = f(p^{r+1}a) \quad \text{and} \quad p^r b \notin T.$$

The mapping g of $\langle S, a\rangle$ (the subgroup of G generated by S and a) onto $\langle T, b\rangle$ given by sending $na + s$ onto $nb + f(s)$ is an isomorphism extending f. It will preserve heights $<\omega\beta$ since if $h_G(na + s) < \omega\beta$ then

$$h_G(na + s) = h_G(s) \qquad \text{and} \qquad h_H(nb + f(s)) = h_H(f(s)) = h_G(s).$$

If

$$h_G(na + s) \geqq \omega\beta \qquad \text{then} \qquad h_G(s) \geqq \omega\beta,$$

so $h_H(f(s)) \geqq \omega\beta$ and hence

$$h_H(nb + f(a)) \geqq \omega\beta.$$

(b) $h_G(a) < \omega\beta$ and for each $m \leqq r$ and each x of the form $p^m a + np^{m+1}a + s$ with $s \in S$, we have $h_G(x) < \omega\beta$. In this case we can successively adjoint $p^r a$, $p^{r-1}a, \cdots, pa$, a just as in Kaplansky's proof, since we are only concerned with heights $<\omega\beta$ where f is height preserving. (In fact f preserves heights $<\omega\beta + \omega$.)

(c) $h_G(a) < \omega\beta$ but for some $m \leqq r$, some x of the form $p^m a + np^{m+1}a + s$ has $h(x) \geqq \omega\beta$. Let x be such an element with m minimal. We first extend f to a g with

$$\text{dom}\,(g) = \langle S, x\rangle = \langle S, p^m a\rangle,$$

as in case (a) getting $g \in I_\beta$. If $m > 0$ we then adjoin $p^{m-1}a, \cdots, pa$, a successively as in case (b). This completes the proof. ⊣

Actually, if one wants to understand the $L_{\infty\omega}$ properties of torsion groups, then it is not desirable to restrict oneself to reduced groups in Theorem 8. For any α, we can define the reduced part of groups G with $l(G) \leqq \alpha$ by a formula $\varphi_\alpha(x)$ of $L_{\infty\omega}$. Thus, if $G \equiv_{\infty\omega} H$ and G is reduced then H is reduced.

On the other hand, there is not a single formula $\varphi(x)$ which works for all p-groups G. Thus, we should really characterize arbitrary

p-groups up to $\equiv_{\infty\omega}$, not just reduced ones. This has been worked out in [1].

It seems clear that a lot of work has been overlooked in algebra just because the notions of $\cong_p$ and especially $\equiv^{\alpha}_{\infty\omega}$ were not known to the workers in the area. For example, one often obtains good results about invariant (i.e., characteristic) subgroups only for countable groups. Perhaps what one should consider are strongly invariant subgroups.

7. METAMATHEMATICAL, PHILOSOPHICAL, HISTORICAL AND BIBLIOGRAPHICAL REMARKS

7.1. METAMATHEMATICAL REMARKS: Let T be some set theory like ZF, Zermelo-Fraenkel, or ZFC, ZF with the axiom of choice. A property $P(x_1 \cdots x_n)$ of sets is said to be *persistent relative to* T if given any models $M = \langle M, E \rangle$ and $M' = \langle M', E' \rangle$ of T, with M' an end extension of M,

$$M \models P[a_1 \cdots a_n] \quad \text{implies} \quad M' \models P[a_1 \cdots a_n],$$

for all $a_1 \cdots a_n, \in M$. (Being an *end* extension of M means that $M \subseteq M'$ and $aE'b$ and $b \in M$ implies $a \in M$ for all $a, b \in M'$.) P is *absolute* relative to T if for all M, M' as above

$$M \models P[a_1 \cdots a_n] \;\text{ iff }\; M' \models P[a_1 \cdots a_n].$$

Intuitively, P is absolute if its meaning does not shift about as you move from one model of set theory to another. (Feferman and Kreisel in [4] have characterized these notions for first order properties P as being Σ_1 relative to T and Δ_1 relative to T.) Now the properties $P_0(f, \mathfrak{A}, \mathfrak{B})$ iff $f\colon \mathfrak{A} \cong \mathfrak{B}$ and $P_1(I, \mathfrak{A}, \mathfrak{B})$ iff $I\colon \mathfrak{A} \cong_p \mathfrak{B}$ are clearly absolute, that is, they have the same meaning in any model of set theory in which f, I, $\mathfrak{A}$, $\mathfrak{B}$ occur. However, $\mathfrak{A} \cong \mathfrak{B}$ is only persistent, *not absolute*. That is, you can have structures $\mathfrak{A}$, $\mathfrak{B}$ in M such that

$$M \models \neg\exists f[f\colon \mathfrak{A} \cong \mathfrak{B}],$$

but

$$M' \models \exists f[f\colon \mathfrak{A} \cong \mathfrak{B}].$$

This can happen even when $\mathfrak{A}$ and $\mathfrak{B}$ have the same cardinality in M. For example, if M is countable and if G and H are groups of power $\aleph_1$ in M with the same Ulm invariants, but which aren't isomorphic then there will be an M' where they are isomorphic, namely, any end extension of M which makes G and H countable.

On the other hand, Theorem 3 shows that $\mathfrak{A} \cong_p \mathfrak{B}$ is invariant by showing that $\mathfrak{A} \not\cong_p \mathfrak{B}$ is persistent. The following is really a meta-theorem.

THEOREM 12: *Let $P(x, y)$ be a first order property absolute relative to T such that*

(1) $\forall \mathfrak{A}, \mathfrak{B}[\mathfrak{A} \cong \mathfrak{B} \rightarrow P(\mathfrak{A}, \mathfrak{B})]$,
is a theorem of T. Then so is

(2) $\forall \mathfrak{A}, \mathfrak{B}[\mathfrak{A} \cong_p \mathfrak{B} \rightarrow P(\mathfrak{A}, \mathfrak{B})]$.
If, in addition to (1),

(3) $\forall \mathfrak{A}, \mathfrak{B}[P(\mathfrak{A}, \mathfrak{B})$ *and* $\mathfrak{A}$ *and* $\mathfrak{B}$ *countable* $\rightarrow \mathfrak{A} \cong \mathfrak{B}]$,
is a theorem of T then so is

(4) $\forall \mathfrak{A}, \mathfrak{B}[\mathfrak{A} \cong_p \mathfrak{B} \leftrightarrow P(\mathfrak{A}, \mathfrak{B})]$.

Proof: Suppose that (1) is a theorem of T and let M be any countable model of T. We need to show $M \models (2)$. So suppose that $\mathfrak{A} \cong_p \mathfrak{B}$ is true in M. By forcing, or by the methods of [2], find an end extension M' of M which is a model of T such that

$$\mathfrak{A} \text{ and } \mathfrak{B} \text{ are countable,}$$

is true in M'. By Theorem 2, and by the absoluteness of $\cong_p$, we have

$$\mathfrak{A} \cong \mathfrak{B},$$

true in M' and hence $P(\mathfrak{A}, \mathfrak{B})$ is true in M', so, by absoluteness, in M. The second part is proved in the same manner. ⊣

Theorem 12 remains true if one weakens the requirement that P be absolute for end extension to the requirement that P be absolute only for Cohen extensions.

We should also point out that the only theorem of this paper which uses some form of the axiom of choice is Theorem 9, because of its dependence on the downward Löwenheim-Skolem lemma of section 2. Other apparent uses (i.e., in Theorem 2) can be easily circumvented.

7.2. PHILOSOPHICAL REMARKS: The metatheorem of section 7.1 shows that any attempt to define an absolute notion of isomorphism which is stronger than $\cong_p$ must fail. This gives strength to the feeling that $\cong_p$ is a very natural notion of isomorphism, one of which mathematicians should be aware. If one proves that $\mathfrak{A} \not\cong \mathfrak{B}$ for specific structures $\mathfrak{A}$ and $\mathfrak{B}$, but leaves open the question

$$\mathfrak{A} \cong_p \mathfrak{B}?$$

then one leaves open the possibility that $\mathfrak{A}$ and $\mathfrak{B}$ are not isomorphic for trivial reasons of cardinality. Or, to put it the other way round, a proof that $\mathfrak{A} \not\cong_p \mathfrak{B}$ is a proof that $\mathfrak{A} \not\cong \mathfrak{B}$ for nontrivial reasons.

7.3. HISTORICAL AND BIBLIOGRAPHICAL REMARKS: The names that need mentioning in addition to those mentioned earlier are Ehrenfeucht and Fraissé, who first published characterizations of elementary equivalence using back and forth arguments. We owe much in our presentation to Chang's proof in [3] of Scott's Theorem. It is from this proof that we have extracted Theorems 6 and 7. Scott's Theorem and the result on invariant relations (Corollary 2 to Theorem 7) were first announced in Scott [11], but the proofs appeared in highly disguised form in Scott [12]. Some of the applications in section **3** first appeared in [1].

We have added a list of suggested reading after the references. We draw special attention to the book by Keisler which gives an up to date treatment of the known model theory for $L_{\omega_1\omega}$ and its sublanguages. It also contains a more extensive bibliography.

Finally, we would like to thank P. Eklof, H. J. Keisler and G. Sabbagh for useful comments on a preprinted version of this article.

REFERENCES

1. Barwise, J., and P. Eklof, "Infinitary properties of Abelian torsion groups," *Ann. Math. Logic*, **2** (1970) 25–68.

2. Barwise, J., "Infinitary methods in the model theory of set theory," *Logic Colloquium '69*, North Holland, 53–66.

3. Chang, C. C., "Some remarks on the model theory of infinitary languages," *The Syntax and Semantics of Infinitary Languages, Lecture Notes in Mathematics*, **72**, Springer-Verlag.

4. Feferman, S., and G. Kreisel, "Persistent and invariant formulas relative to theories of higher type," *Bull. Amer. Math. Soc.*, **72** (1966), 480–485.

5. Fuchs, L., *Abelian Groups*. New York: Pergamon Press, 1960.

6. Halmos, P., *Naive Set Theory*. Princeton, N.J.: Van Nostrand, 1960.

7. Kaplansky, I., *Infinite Abelian Groups*. Ann Arbor: University of Michigan Press, 1968.

8. Karp, C., "Finite quantifier equivalence," *The Theory of Models*. Amsterdam: North Holland, 1965, 407–412.

9. Kueker, D., "Definability, automorphisms and infinitary languages," *The Syntax and Semantics of Infinitary Languages, Lecture Notes in Mathematics*, **72**, Springer-Verlag.

10. Moschovakis, Y., "Abstract first order computability II," *Trans. Amer. Math. Soc.*, **138** (1969), 465–504.

11. Scott, D., "Logic with denumerably long formulas and finite strings of quantifiers," *Theory of Models*. Amsterdam: North Holland, 1965, 329–341.

12. Scott, D., "Invariant borel sets," *Fund. Math.*, **56**, 117–128.

Further reading:

13. Barwise, J., "Abstract logics and $L_{\infty\omega}$," *Ann. of Math. Logic*, **4** (1972), 309–340.

14. Benda, M., "Reduced products and nonstandard [infinitary] logics," *J. Symbolic Logic*, **34** (1968), 424–438.

15. Eklof, P., and G. Sabbagh, "Definability problems for modules and rings," *J. Symbolic Logic*, **36** (1971), 623–649.

16. Feferman, S., "Infinitary properties, local functors and systems of ordinal functions," *Conference in Mathematical Logic—London '70, Lecture Notes in Mathematics*, 255, Springer-Verlag, 63–97.

17. Keisler, H. J., "Formulas with linearly ordered quantifiers," *The Syntax and Semantics of Infinitary Languages, Lecture Notes in Mathematics*, **72**. Springer-Verlag, 96–130.

18. ———, *Model Theory of Infinitary Languages*. Amsterdam: North Holland, 1971.

19. Morley, M., "The number of countable models," *J. Symbolic Logic*, **35** (1970), 14–18.

20. Eklof, P. C., "Infinitary equivalence of Abelian groups," *Fund. Math.*, to appear.

NON-STANDARD ANALYSIS

Allen R. Bernstein

INTRODUCTION

The existence of non-standard models has been known for quite some time. The term itself first appeared in a 1934 paper of Skolem [17] in which he constructed a non-standard model (in modern parlance a proper elementary extension) of the natural numbers. In fact the construction he used bears a resemblance to the ultra-product construction discovered by Łoś later in 1954 [9] and used in this paper.

With the extension to uncountable languages of the Gödel completeness theorem and corollary compactness theorem ([7], [11], [13]) it became clear that any infinite model possesses a proper elementary extension. These results found application in the 1950's to a number of problems chiefly of an algebraic flavor. Then in 1961 [13] Abraham Robinson showed that one could do a very interesting thing with elementary extensions of the real numbers. First of all these extensions are ordered field extensions of the real numbers and consequently must have infinitesimal as well as infinite elements. Then using these infinitesimal elements one could develop the theory of the calculus in a manner almost identical to that suggested by Leibnitz in the seventeenth century. This development is carried out in [13] and more extensively in [16].

After restoring the previously discredited methods of Leibnitz regarding the calculus it was natural for Robinson to seek to apply his methods to solve new problems in analysis. However the non-standard calculus was developed using only the results available for first order model theory while most of the notions of current day mathematics deal with higher order concepts. Furthermore there is no completeness or compactness result for higher order languages comparable in strength to the first order theory. In fact, as is well known, there is no other model which satisfies the same second order sentences as the natural numbers.

However there is a weaker result which Robinson realized could be used to handle higher order problems. Henkin showed in [8] that if the definition of truth in higher order models were weakened from the natural definition then in fact the completeness theorem would be true so that non-standard models could be obtained. What Henkin's idea amounted to was taking non-standard models of the underlying set theory.

The first results using this higher order theory were obtained in [14] where Robinson developed a theory of non-standard complex numbers and used this theory to obtain some new standard results concerning zeros of complex polynomials. Further development during the early 1960's included topics in function theory, topology, functional analysis, topological groups, etc., and are described in Robinson's book [16]. An excellent compendium of later work by various authors may be found in [10].

This paper will focus on a particular application of non-standard analysis which was used to solve an open question in functional analysis. The question involves the long standing still open problem of whether or not every bounded linear operator on Hilbert space (or even Banach space) possesses a proper invariant subspace. For compact operators this was proved in the affirmative for Hilbert space by N. Aronszajn and J. von Neumann in the 1930's and for Banach space by N. Aronszajn and K. Smith in 1954 [1]. In a survey article in 1963 [6], P. Halmos listed the following open question of K. Smith: If T is a bounded linear operator on a Hilbert space such that T^2 is compact, then does T have a proper invariant subspace?

This was answered in the affirmative in [4] where it was shown that if some polynomial of T is compact then T has a proper invariant subspace. This result was extended to Banach space in [3].

In sections 3–6 we present essentially the same proof given in [4]. Sections 1 and 2 contain a development of the theory of non-standard analysis leading up to this result. There are several basic settings in which it is possible to carry out this development. One way is to consider non-standard models of set theory (see [12]). Another way which we adopt here is to keep the set theory fixed and change the definition of model. This is the approach used by Henkin in [8] and Robinson in [16] and has the advantage of being more accessible to those not accustomed to dealing with a shift in set theory.

The development here however is somewhat different than in the preceding two references. There the higher order models were obtained by altering a suitable first order model. The approach here is to define the notion of a *higher-order ultraproduct* and use this to construct the non-standard models directly.

1. HIGHER ORDER MODEL THEORY

In this section we describe a higher order language L and a suitable corresponding model theory.

The set $\mathscr{T}$ of *types* is defined to be the smallest set satisfying:

(i) $0 \in \mathscr{T}$.
(ii) If $\tau_1, \cdots, \tau_n \in \mathscr{T}$ then the finite sequence $(\tau_1, \tau_2, \cdots, \tau_n) \in \mathscr{T}$.

The *alphabet* of L consists of the following disjoint sets of symbols:

(i) For each $\tau \in \mathscr{T}$, a set C_τ of constant symbols of type τ.
(ii) For each $\tau \in \mathscr{T}$, a countable set $\{v_{\tau 1}, v_{\tau 2}, \cdots, v_{\tau n}, \cdots\}$ of variables of type τ.
(iii) Logical symbols: $\exists$, $\neg$, $\wedge$, ( ), $\equiv$.

The *atomic formulas* are of one of the following two forms:

(i) $\alpha \equiv \beta$ where α and β are constants or variables (1 of each permitted) of the same type.
(ii) $\alpha(\beta_1, \beta_2, \cdots, \beta_n)$ where α is a constant or variable of type $\tau = (\tau_1, \tau_2, \cdots, \tau_n)$ and $\beta_1, \beta_2, \cdots, \beta_n$ are constants or variables (some of each permitted) of types $\tau_1, \tau_2, \cdots, \tau_n$ respectively.

The set $\mathscr{F}$ of *formulas* is defined to be the smallest set satisfying:

(i) If ϕ is an atomic formula then $\phi \in \mathscr{F}$.
(ii) If $\phi \in \mathscr{F}$ then $\neg\phi \in \mathscr{F}$.
(iii) If $\phi, \psi \in \mathscr{F}$ then $(\phi \wedge \psi) \in \mathscr{F}$.
(iv) If $\phi \in \mathscr{F}$ and x is a variable then $\exists x\phi \in \mathscr{F}$.

We make the customary abbreviations $\forall$, $\vee$, $\rightarrow$, $\leftrightarrow$ and define the notion of a *free occurrence* of a variable as usual. Then a *sentence* is a formula in which no variable occurs free.

If we have a formula with free variables $x_1, \cdots, x_n$ we may denote such a formula by $\phi(x_1, \cdots, x_n)$ or $\psi(x_1, \cdots, x_n)$, etc. Then if $\alpha_1, \cdots, \alpha_n$ are variables or constants of the same type as $x_1, \cdots, x_n$ respectively, then $\phi[\alpha_1, \cdots, \alpha_n]$ denotes the formula obtained from $\phi(x_1, \cdots, x_n)$ by substituting for each i all free occurrences of x_i by α_i.

Next we wish to define the notion of *model* relative to a given higher order language L.

DEFINITION: A *structure* is a collection $\{A_\tau\}_{\tau \in \mathscr{T}}$ of sets where each $A_\tau \neq \phi$ and for $\tau = (\tau_1, \tau_2, \cdots, \tau_n)$, A_τ is a set of subsets of $A_{\tau_1} \times A_{\tau_2} \times \cdots \times A_{\tau_n}$.

DEFINITION: A *model* is a pair $M = \langle \{A_\tau\}_{\tau \in \mathscr{T}}, F \rangle$ where $\{A_\tau\}_{\tau \in \mathscr{T}}$ is a structure and F is a function on $\bigcup_{\tau \in \mathscr{T}} C_\tau$ s.t. F maps C_τ into A_τ.

In the above it is understood that if $\tau = (\tau_1)$ then the one-termed Cartesian product is just A_{τ_1} so that A_τ is a set of subsets of A_{τ_1}. If $a \in A_0$ then a is called an *individual* of M and if $Q \in A_\tau$, $\tau \neq 0$, then Q is called a *relation* of M.

Let $M = \langle \{A_\tau\}_{\tau \in \mathcal{T}}, F \rangle$ be a model. Given a sentence ϕ of the corresponding language L we wish to define when ϕ is *true* in M, written $M \models \phi$. This is done by induction on ϕ. First suppose F is onto $\bigcup_{\tau \in \mathcal{T}} A_\tau$.

(i) For an atomic sentence ϕ,

(a) If ϕ is $\alpha \equiv \beta$ then $M \models \phi$ iff $F(\alpha) = F(\beta)$.

(b) If ϕ is $\alpha(\beta_1, \cdots, \beta_n)$ then $M \models \phi$ iff

$$\langle F(\beta_1), \cdots, F(\beta_n) \rangle \in F(\alpha).$$

(ii) $M \models \neg\phi$ iff not $M \models \phi$.

(iii) $M \models (\phi \wedge \psi)$ iff $M \models \phi$ and $M \models \psi$.

(iv) $M \models \exists x \phi(x)$ iff $M \models \phi[c]$ for some constant symbol c.

If F is not onto $\bigcup_{\tau \in \mathcal{T}} A_\tau$ then extend F to F' which is (extending the language L to L' appropriately) and define

$$\langle \{A_\tau\}_{\tau \in \mathcal{T}}, F \rangle \models \phi \quad \text{iff} \quad \langle \{A_\tau\}_{\tau \in \mathcal{T}}, F' \rangle \models \phi.$$

It is not difficult to verify that the preceding definition is legitimate, i.e., independent of the particular extension F' of F. Observe that the above definition does not interpret higher order statements in a completely literal fashion. For example, suppose we have a sentence of the form $\forall x \phi(x)$ where say x is a variable of type (0). Then this sentence will be true in M iff S satisfies $\phi(x)$ for all $S \in A_{(0)}$. Thus we exclude from consideration all subsets of A_0 which are not in $A_{(0)}$.

Finally we introduce the following notation. Suppose $M = \langle \{A_\tau\}, F \rangle$ is a model for L and suppose $\phi(x_1, \cdots, x_n)$ denotes a formula of L with free variables among $x_1, \cdots, x_n$. Let $a_1, \cdots, a_n \in \bigcup A_\tau$. Then we write

$$M \models \phi[\![a_1, \cdots, a_n]\!],$$

iff $M' \models \phi[\alpha_1, \cdots, \alpha_n]$ where $M' = \langle \{A_\tau\}, F' \rangle$ with $F' \supseteq F$ and $F'(\alpha_i) = a_i$, $i = 1, 2, \cdots, n$.

2. HIGHER ORDER ULTRAPRODUCTS

Let I be a non-empty set. Then a collection D of subsets of I is said to be an ultrafilter on I if

(1) $\phi \neq D$, $\phi \notin D$,
(2) $X, Y \in D \Rightarrow X \cap Y \in D$,
(3) $X \in D, X \subseteq Y \subseteq I \Rightarrow Y \in I$, and
(4) $X \cup Y \in D \Rightarrow X \in D$ or $Y \in D$.

D is a *principal* ultrafilter if $\bigcap D \in D$, otherwise *non-principal*. There exist non-principal ultrafilters on any infinite set I and such ultrafilters must of necessity contain only infinite sets.

Now let I be a non-empty set and let $\{M_i\}_{i \in I}$ be a collection of higher order models where

$$M_i = \langle \{A_\tau^{(i)}\}_{\tau \in \mathcal{T}}, F_i \rangle \qquad \text{for each } i \in I.$$

For any $\tau \in \mathcal{T}$ denote by π_τ the Cartesian product,

$$\pi_\tau = \prod_{i \in I} A_\tau^{(i)} = \{f \mid D(f) = I \quad \text{and} \quad f(i) \in A_\tau^{(i)} \quad \text{all} \quad i \in I\}.$$

For each $f \in \pi_\tau$ we wish to define a corresponding $\bar{f}$ as follows by induction on τ.

(2.1) If $\tau = 0$ then for $f \in \pi_0$, we define $\bar{f} = \{g \mid g \sim f\}$, where $g \sim f \Leftrightarrow \{i \mid f(i) = g(i)\} \in D$. Thus

$$\bar{f} = \bar{g} \Leftrightarrow \{i \mid f(i) = g(i)\} \in D.$$

(2.2) If $\tau = (\tau_1, \cdots, \tau_n)$ then for $g \in \pi_\tau$, $f_k \in \pi_{\tau_k}$, $k = 1, \cdots, n$,

$$\langle \bar{f}_1, \cdots, \bar{f}_n \rangle \in \bar{g} \Leftrightarrow \{i \mid \langle f_1(i), \cdots, f_n(i) \rangle \in g(i)\} \in D.$$

We must verify that the above is well-defined, i.e., that if $h_k \in \pi_{\tau_k}$, $\bar{h}_k = \bar{f}_k$, $k = 1, \cdots, n$, then

$$\text{(2.3)} \qquad \{i \mid \langle f_1(i), \cdots, f_n(i) \rangle \in g(i)\} \in D$$
$$\Leftrightarrow \{i \mid \langle h_1(i), \cdots, h_n(i) \rangle \in g(i)\} \in D.$$

Before verifying (2.3) we wish to show that if in fact $\bar{g}$ is well defined for all $g \in \pi_\tau$ by (2.2) then

$$\text{(2.4)} \qquad \text{For } g, h \in \pi_\tau, \qquad \bar{g} = \bar{h} \Leftrightarrow \{i \mid g(i) = h(i)\} \in D.$$

To verify (2.4) note that the right side implies the left side, as follows easily from (2.2), and the fact that

$$\{i \mid \langle f_1(i), \cdots, f_n(i) \rangle \in h(i)\} \supseteq \{i \mid g(i) = h(i)\}$$
$$\cap \{i \mid \langle f_1(i), \cdots, f_n(i) \rangle \in g(i)\}.$$

To go the other way suppose that $\{i \mid g(i) = h(i)\} \notin D$. Then $S = \{i \mid g(i) \neq h(i)\} \in D$. Let $S = S_1 \cup S_2$ where

$$S_1 = \{i \mid g(i) - h(i) \neq \phi\}$$

$$S_2 = \{i \mid h(i) - g(i) \neq \phi\}.$$

Then, since $S \in D$, either $S_1 \in D$ or $S_2 \in D$, say the former. For each $i \in S_1$ let $\langle a_{i1}, \cdots, a_{in}\rangle \in g(i) - h(i)$. Then define

$$f_k \in \pi_{\tau k}, \qquad k = 1, \cdots, n,$$

by letting

$$f_k(i) = a_{ik} \qquad \text{for } i \in S_1.$$

$$f_k(i) = \text{anything} \qquad \text{for } i \notin S_1.$$

$\{i \mid \langle f_1(i), \cdots, f_n(i)\rangle \in g(i)\} \supseteq S_1$, so is in D.

$\{i \mid \langle f_1(i), \cdots, f_n(i)\rangle \in h(i)\} \supseteq I - S_1$, so is not in D.

Thus, since we are assuming (2.2) is well-defined for $g, h \in \pi_\tau$, then $\langle \bar{f}_1, \cdots, \bar{f}_n\rangle \in \bar{g} - \bar{h}$. Thus $\bar{g} \neq \bar{h}$ completing the verification of (2.4).

Now we are ready to verify (2.1) by induction on τ. Observe that things are well-defined for $\tau = 0$ so we let $\tau = (\tau_1, \cdots, \tau_n)$ and suppose $\bar{f}_k$ is well-defined for elements f_k of $\pi_{\tau k}$, $k = 1, \cdots, n$. By the symmetry of the situation we need only show the implication in one direction in (2.3). Thus let $g \in \pi_\tau$, $f_k, h_k \in \pi_{\tau k}$, $\bar{h}_k = \bar{f}_k$, $k = 1, \cdots, n$. Then

$$\{i \mid \langle h_1(i), \cdots, h_n(i)\rangle \in g(i)\} \supseteq \{i \mid \langle f_1(i), \cdots, f_n(i)\rangle \in g(i)\}$$
$$\cap \{i \mid h_1(i) = f_1(i)\} \cap \cdots \cap \{i \mid h_n(i) = f_n(i)\}.$$

Therefore, by using the left side of (2.3), the induction hypothesis together with (2.4), and the properties of D, we get the right side of (2.3).

Now suppose L is a higher order language and $\{M_i\}_{i \in I}$ is a collection of models for L with

$$M_i = \langle \{A_\tau^{(i)}\}_{\tau \in \mathcal{T}}, F_i\rangle \qquad \text{each } i \in I.$$

We define $M = D$-product M_i as follows:

$$M = \langle \{A_\tau\}_{\tau \in \mathcal{T}}, F\rangle,$$

where

(2.5) For each $\tau \in \mathscr{T}$, $A_\tau = \{\bar{f} \mid f \in \pi_\tau\}$.

(2.6) For any constant α of L, $F(\alpha) = \bar{f}$, where f is defined by

$$f(i) = F_i(\alpha), \qquad i \in I.$$

With the above definition of D-product it is possible to prove the following fundamental theorem. The proof is virtually identical to that of the first order version of the same theorem (cf. [2] p. 90) and will not be given here.

Łoś' Theorem (Higher order version): *Let M = D-product M_i as above. Let $\phi(x_1, \cdots, x_n)$ be a formula of L with free variables among $x_1, \cdots, x_n$.* Let $\bar{f}_1, \cdots, \bar{f}_n \in \bigcup_{\tau \in \mathscr{T}} A_\tau$. *Then*

$$M \models \phi[\![\bar{f}_1, \cdots, \bar{f}_n]\!] \quad \textit{iff} \quad \{i \mid M_i \models \phi[\![f_1(i), \cdots, f_n(i)]\!]\} \in D.$$

Next consider the case where all of the M_i are the same model for L, say N where $N = \langle\{B_\tau\}_{\tau \in \mathscr{T}}, G\rangle$. In this case we write M = D-power N. Write $M = \langle\{A_\tau\}_{\tau \in \mathscr{T}}, F\rangle$. By definition of M, for each constant α of L, $F(\alpha) = \bar{f}_{G(\alpha)}$ where $f_{G(\alpha)}$ is the constant function, $f(i) = G(\alpha)$, $i \in I$. It follows from Łoś' Theorem that for any sentence ϕ of L,

$$(2.7) \qquad M \models \phi \quad \text{iff} \quad N \models \phi.$$

For $a \in B_0$ we may identify the equivalence class $\bar{f}_a$ of the constant function $f_a(i) = a$, all $i \in I$, by a itself so that

$$(2.8) \qquad B_0 \subseteq A_0 \qquad \text{and} \qquad F(\alpha) = G(\alpha) \qquad \text{for all } \alpha \in C_0.$$

We call an M satisfying (2.7) and (2.8) an *L-extension* of N.

Observe that if B_0 is infinite then we may always choose an I and a D so that M turns out to be a proper L-extension of N, in particular A_0 properly includes B_0. To do this let I be a set with cardinality equal to that of B_0 and let f be a one-one function from I onto B_0. If D is a non-principal ultrafilter on I then, since D contains only infinite sets, $\{i \mid f(i) = b\} \notin D$ for any $b \in B_0$. Thus $\bar{f} \in A_0 - B_0$.

3. NON-STANDARD HILBERT SPACE

We now wish to apply the results of the preceding section to the theory of Hilbert space. To this end let H be the set of elements of a Hilbert space over the set of complex numbers C. If H is non-separable then given any bounded linear operator T and any non-zero $x \in H$ the set $\{x, Tx, T^2x, \cdots\}$ generates a proper invariant subspace for H. Therefore we shall suppose hereafter that H is separable.

Let $A_0 = H \cup C$ and for each type τ of the form $\tau = (\tau_1, \cdots, \tau_n)$ let A_τ be the set of *all* subsets of $A_{\tau_1} \times \cdots \times A_{\tau_n}$. Then let $S = \{A_\tau\}_{\tau \in \mathscr{T}}$ so S is the largest structure having A_0 as its set of individuals. Thus among the relations of S are H and C together with all their subsets, relations, relations of relations, etc. In particular the set of real numbers R and the set of positive integers Z, being subsets of C, will be relations of S. Furthermore, we may single out the relations under which H is a Hilbert space over the complex numbers C. Thus for example, we can find in M the relation $Q(z, y, z)$ which holds just in case $x, y, z \in C$ and $x - y = z$ and the relation $A(x, y, z)$ which holds just in case $x, y, z \in H$ and $x + y = z$. In this manner all the algebraic operations over H and C are relations of S. Similarly we can find in S the relation $N(x, y)$ which holds just in case $x \in H$, $y \in R$ and the norm of x, $\|x\|$, is equal to y. A function of n variables in this system is a relation of $n + 1$ variables in which the first n coordinates determine the $(n + 1)$-st coordinate uniquely. Thus, for example, a sequence of elements of H, $\{x_n\}$, which is a function from Z into H, is a two-placed relation $P(n, y)$ which holds just in case $y = x_n$.

Next we choose a suitable higher-order language L to correspond to S. Thus we take disjoint sets of constant symbols C_τ, $\tau \in \mathscr{T}$, with 1-1 functions F_τ from C_τ onto A_τ. Then let $F = \bigcup_{\tau \in \mathscr{T}} F_\tau$ and $M = \langle\{A_\tau\}_{\tau \in \mathscr{T}}, F\rangle$. Hence the language L has a distinct constant symbol for every individual and relation of M.

Now let $^*M = \langle\{B_\tau\}_{\tau \in \mathscr{T}}, G\rangle$ be .a proper L-extension of M. Consider an arbitrary relation Q of M, $Q \in A_\tau$. Q is denoted by some constant symbol β of L, and β must denote in $^*M$ some element of B_τ which we write as $^*Q$. $^*Q$ will have the same properties as Q to the extent to which these can be expressed as sentences of L. In particular H, C, R, and Z will extend in $*M$ to the sets $*H$, $*C$, $*R$, $*Z$ which properly include H, C, R, and Z respectively. Thus the non-standard Hilbert space $*H$ will contain in addition to all standard points of H, other non-standard points which are not elements of H. Similarly $*Z$ will contain elements which do not belong to Z and these are called infinite integers. If $Q \in A_\tau$, $\tau \neq 0$, then Q is called an *internal* relation. In general there will be relations built up from A_0 which are not internal.

Any function f which is a relation of M corresponds to a relation $*f$ of $*M$ which again must be a function since this fact can be expressed as a sentence of L. If, in addition, f is a function from individuals to individuals, then $*f$ will be an extension of f. Thus for example a sequence $\{x_n\}$ of elements of H (or C) which is a function f from Z into H (or C), extends to a "sequence" $*\{x_n\}$ which is just the function $*f$ defined now for all $n \in *Z$ and taking values in $*H$ (or $*C$). If $m \in *Z$, then x_m will designate the element of $*H$ (or $*C$) corresponding to m in $*\{x_n\}$, i.e., $x_m = *f(m)$. Since $*f(m) = f(m)$ for all $m \in Z$, this designation is unambiguous.

Let Σ be the set of sequences in M. Its extension to $*M$, $*\Sigma$, will contain all standard sequences, that is all extensions $*\{x_n\}$ of sequences $\{x_n\}$ in Σ. In addition, $*\Sigma$ will contain other elements which must be functions from $*Z$ into $*A_0$ since this fact can be expressed as a sentence of L. Any element $\{t_n\}$ of $*\Sigma$ we call a *sequence.

For the algebraic operations on H and C we use the same symbol for their extensions to $*M$. Thus for example the extension of the relation of addition, $+$, for complex numbers is in $*M$ a relation over elements of $*C$ and will again be denoted by "$+$". Likewise the norm $\|\cdot\|$ which in M is a function from H into R extends to a function from $*H$ into $*R$ which is again denoted by "$\|\cdot\|$".

Consider an arbitrary element $r \in *R$. If $|r| < s$ for all $s \in R$, $s > 0$, then r is called *infinitesimal*. If $|r| < s$ for some $s \in R$ then r is called *finite*. In this case there is a unique element of R, denoted by $°r$, such that $r - °r$ is infinitesimal. This last fact follows easily from the completeness of R and is proved by defining $°r$ as the real number determined by the Dedekind cut $\langle\{s \in R \mid |s| < r\},$

$\{s \in R \mid |s| \geqq r\}\rangle$ (cf. [16]). We write $r_1 \sim r_2$ if $r_1 - r_2$ is infinitesimal and it is easily verified that this is an equivalence relation. We make the same definitions for the elements of $^*C$. Thus an element c of $^*C$ is called infinitesimal, finite, or infinite according to what $|c|$ is. Once again, for finite c there is a unique $^\circ c \in C$ such that $c - {}^\circ c$ is infinitesimal, $c \sim {}^\circ c$.

THEOREM 3.1: *The sequence $\{s_n\}$ of standard real numbers converges to the standard real number s, $\lim_{n\to\infty} s_n = s$, iff $s - s_\nu$ is infinitesimal for all infinite integers ν.*

Proof: Suppose $\lim s_n = s$. Let ν be an infinite integer so $\nu \in {}^*Z - Z$ and let $\varepsilon \in R$, $\varepsilon > 0$. Then there is a $k \in Z$, such that the following statement is true about R:

$$\forall m(m \in Z \wedge m \geqq k \rightarrow |s_m - s| < \varepsilon).$$

The above statement can be translated into a sentence of L whose truth in M implies its truth in $^*M$. The corresponding statement about $^*R$ says:

$$\forall m(m \in {}^*Z \wedge m \geqq k \rightarrow |s_m - s| < \varepsilon).$$

In particular, since $\nu \geqq k$, we have $|s_\nu - s| < \varepsilon$. Since ε was an arbitrary positive element of R we have shown that $s_\nu - s$ is infinitesimal. The proof of the converse will be left as an exercise and will not be used here.

Let x be an element of $^*H$. If $x \in H$ we say x is standard. If $\|x\| < r$ for some $r \in R$, x is called norm-finite. If $\|x\| < r$ for all positive $r \in R$, then x is called infinitesimal. We write $x \sim y$ for two points $x, y \in {}^*H$ such that $x - y$ is infinitesimal and this is an equivalence relation over $^*H$. It is easy to verify that if $x_1 \sim y_1$ and $x_2 \sim y_2$ then $x_1 + x_2 \sim y_1 + y_2$ and that $cx_1 \sim cy_1$ providing c is a finite element of $^*C$. If there is a standard $y \in {}^*H$ (i.e., $y \in H$) such that $x \sim y$ then x is called near-standard. In this case there is a unique such y and we write $y = {}^\circ x$. $^*H$ will contain points which are norm-finite but not near-standard.

THEOREM 3.2: *Let $x \in {}^*H$ be not near-standard. Then there is a standard $\varepsilon > 0$ such that $\|x - b\| > \varepsilon$ for all standard b in $^*H$.*

Proof: Suppose that the conclusion of the theorem is false. For each $k \in Z$, we may therefore pick an element b_k in H such that $\|x - b_k\| < 1/2k$. Then if $m, n > k$,

$$\|b_m - b_n\| \leqq \|b_m - x\| + \|x - b_n\| < \frac{1}{2m} + \frac{1}{2n} < \frac{1}{2k} + \frac{1}{2k} = \frac{1}{k}.$$

Thus $\{b_n\}$ is a Cauchy sequence and converges to some point b in H. Now choose any standard $\varepsilon > 0$, and let $n \in Z$ be greater than $1/\varepsilon$ and such that $\|b - b_n\| < \varepsilon/2$. Then

$$\|x - b\| \leqq \|x - b_n\| + \|b_n - b\| < \frac{\varepsilon}{2} + \frac{\varepsilon}{2} = \varepsilon.$$

This shows that $\|x - b\|$ is less than any standard $\varepsilon > 0$ and must therefore be infinitesimal. Since b is standard, x is near-standard and this contradiction proves the theorem.

THEOREM 3.3: *Let S be a compact set of points of H. Then every point of $^*S$ is near-standard.*

Proof: Suppose there is a point $x \in {}^*S$ which is not near-standard. Using the previous theorem, let ε be a standard positive number such that $\|x - b\| > \varepsilon$ for all standard points b. Since S is compact it contains standard points $b_1, b_2, \cdots, b_n$, $n \in Z$, such that it is true in M that "for every point ξ in S at least one of the numbers $\|\xi - b_1\|$, $\|\xi - b_2\|, \cdots, \|\xi - b_n\|$ is smaller than ε". (We are excluding the trivial case where S is empty.) The statement in quotes may be expressed as a sentence of L and therefore is true also in $^*M$. Specifying ξ to x, we conclude that $\|x - b_i\| < \varepsilon$ for some i, $1 \leqq i \leqq n$. This contradicts our assumption that $\|x - b\| > \varepsilon$ for all standard points b and completes the proof.

Let V be a relation of M of type (τ) so that V is a set consisting of elements of type τ. If the elements of V are described by some common mathematical name then this name may be used again to refer to the elements of $^*V$. For example, if V is the set of bounded operators on H, the set of projections on H, or the set of linear subspaces of H, then the elements of $^*V$ will be referred to as "bounded operators on $^*H$", "projections on $^*H$", or "linear subspaces of $^*H$".

Let W be the relation of M which is the set of all the bounded linear operators on H. There is a real-valued norm function, $\|\cdot\|$, defined on elements $T \in W$ by

$$\|T\| = \sup_{\|x\|=1} \|Tx\|.$$

As we pass to $^*M$, this function extends to a function from $^*W$ (the set of bounded linear operators on $^*H$) to $^*R$ which will again be denoted by $\|\cdot\|$ and which has the property that $\|Tx\| \leqq \|T\|\ \|x\|$ for all $T \in {}^*W$ and all $x \in {}^*H$. For $T \in W$, we denote its extension to an operator on $^*H$ by $^*T$ as usual.

THEOREM 3.4: *Let T be a bounded linear operator on H. Then $^*T$ transforms every norm-finite point of $^*H$ into a norm-finite point.*

Proof: If x is a norm-finite point of $^*H$, then

$$\|{}^*Tx\| \leqq \|{}^*T\|\ \|x\| = \|T\|\ \|x\|,$$

since $\|{}^*T\| = \|T\|$. Thus $\|{}^*Tx\|$ is less than the finite element of $^*R$, $\|T\|\ \|x\|$, and hence $^*Tx$ is norm-finite.

A bounded linear operator T on H is called compact (or completely continuous) if it transforms every bounded set of points of H into a conditionally compact set, i.e., a set whose closure is compact. Finally we have the following fundamental characterization of compact operators given by Robinson.

THEOREM 3.5: *Let T be a compact operator on H. Then $^*Tx$ is near-standard for all norm-finite x in $^*H$.*

Proof: If x is norm-finite then $\|x\| < r$ for some standard $r \in R$. The set $D = \{\xi \mid \|\xi\| < r\}$ in B is bounded and is therefore mapped by T on a set whose closure, A, is compact. If the corresponding sets in $^*H$ are $^*D$ and $^*A$ respectively then $^*D$ contains x (since x satisfies the defining condition of D) and so $^*A$ contains $^*Tx$. But $^*A$ contains only near-standard points, by 3.3, so $^*Tx$ is near-standard, proving Theorem 3.5.

Now let $\{e_i\}$, $i \in Z$, be an orthonormal basis for H. Denote by H_n the space spanned by $\{e_1, \cdots, e_n\}$ and by P_n the projection of H onto H_n.

As we pass to $^*M$, $\{e_i\}$ is extended to a *sequence $^*\{e_i\}$ of elements of $^*H$, where i now ranges over $^*Z$. If ν is an infinite positive integer, then H_ν will designate the subspace of $^*H$ spanned by $\{e_1, e_2, \cdots, e_\nu\}$. H_ν consists of all elements of $^*H$ which can be expressed in the form $\sum_{i=1}^{\nu} a_i e_i$ with $a_i \in {}^*C$, $i = 1, 2, \cdots, \nu$. However, this summation sign cannot now be regarded as representing the iteration of the addition operation in the ordinary manner, but must be regarded as defined by transfer from M. For this purpose let Δ be the set of finite sequences of elements of H. Δ extends to $^*\Delta$ which consists of all internal functions from some initial segment of $^*Z$ to $^*H$. Then the function Σ which assigns to each element of Δ its sum has an extension $^*\Sigma$ which assigns to each element of $^*\Delta$ an element of $^*H$ which is its "sum". For $\nu \in {}^*Z$ we use the usual notation $\sum_{i=1}^{\nu} \xi_i$ to denote the value of $^*\Sigma$ applied to the element $\{\xi_1, \cdots, \xi_\nu\}$ of $^*\Delta$. The notion of sum for infinite sequences likewise extends to $^*M$. Thus for a *sequence of elements of $^*H$, $\{\xi_i\}_{i \in {}^*Z}$, we write $\sum_{i=1}^{\infty} \xi_i = \sigma$ to mean that given any $\varepsilon \in {}^*R$, $\varepsilon > 0$, there is a $\nu \in {}^*Z$ such that $|\sum_{i=1}^{\nu} \xi_i - \sigma| < \varepsilon$ for all $\mu > \nu$, $\mu \in {}^*Z$. We can deal similarly with sequences of elements of $^*C$.

THEOREM 3.6: *If $x \in H$ then $\|x - P_\nu x\|$ is infinitesimal for all infinite integers ν.*

Proof: Let $x \in H$. Then since $\{e_i\}$ is a basis for H, $\lim_{n\to\infty} \{P_n x\} = x$. It follows from Theorem 3.1 that for any infinite integer ν the νth term of the sequence $^*\{P_n x\}$, namely $P_\nu x$, is infinitesimally close to x.

THEOREM 3.7: *If x is a near-standard element of $^*H$ then x is norm-finite and $\|x - P_\nu x\|$ is infinitesimal.*

Proof: Let $x \in H$ and suppose $\|x - {}^\circ x\|$ is infinitesimal for some ${}^\circ x \in H$. Then by the triangle inequality,

$$\|x\| \leqq \|x - {}^\circ x\| + \|{}^\circ x\| \leqq 1 + \|{}^\circ x\|,$$

which shows that x is norm-finite. Furthermore,

$$\|x - P_\nu x\| \leqq \|x - {}^\circ x\| + \|{}^\circ x - P_\nu {}^\circ x\| + \|P_\nu\| \, \|{}^\circ x - x\|.$$

Thus it follows from Theorem 3.7 and the fact that $\|P_\nu\| = 1$ that $\|x - P_\nu x\|$ is infinitesimal.

4. INFINITE MATRICES

Let T be a bounded linear operator on H. Then relative to an orthonormal basis $\{e_i\}$ T is represented by an infinite matrix $[a_{jk}]$, $j, k \in Z$. Thus for any $k \in Z$, $Te_k = \sum_{j=1}^{\infty} a_{jk}e_j$. Since $[a_{jk}]$ is just a function from $Z \times Z$ into C it has an extension $^*[a_{jk}]$ which is a function from $^*Z \times {}^*Z$ into $^*C$. For $m, n \in {}^*Z$ the value of this function at $\langle m, n\rangle$ is denoted by a_{mn} as usual.

THEOREM 4.1: *Let T be a compact operator represented by $[a_{jk}]$. Then a_{jk} is infinitesimal for infinite j, k finite or infinite.*

Proof: For any $k \in {}^*Z$ consider the point $^*Te_k = \sum_{i=1}^{\infty} a_{ik}e_i$. Since e_k is norm-finite and T is compact it follows from Theorem 3.5 that $^*Te_k$ is infinitesimally close to a standard point $y \in H$. Write $y = \sum_{i=1}^{\infty} y_ie_i$ so in particular $\lim_{i\to\infty} y_i = 0$. Then

$$\|{}^*Te_k - y\|^2 = \|\sum_{i=1}^{\infty} (a_{ik} - y_i)e_i\|^2 = \sum_{i=1}^{\infty} (a_{ik} - y_i)^2 = \eta \sim 0.$$

Now let j be an infinite integer. From the preceding equality we see that $(a_{jk} - y_j)^2 < \eta \sim 0$, therefore $|a_{jk} - y_j| < \sqrt{\eta} \sim 0$. But since $\lim_{i\to\infty} y_i = 0$ it follows from Theorem 3.1 that $y_j \sim 0$. Thus $|a_{jk}| < \sqrt{\eta} + |y_j| \sim 0$ so that a_{jk} is infinitesimal completing the proof.

Given any infinite matrix $[a_{jk}]$ we may multiply it by itself n times in the usual manner obtaining a new matrix $[a_{jk}]^n$. The terms of this matrix are denoted by $a_{jk}^{(n)}$, thus $[a_{jk}]^n = [a_{jk}^{(n)}]$.

A matrix $[a_{jk}]$ is called *almost superdiagonal* if $a_{jk} = 0$ for $j > k + 1$.

THEOREM 4.2: *Let $[a_{jk}]$ be an almost superdiagonal matrix. Then for any positive integers, s, m, n*

(i) $a_{n+m,n}^{(s)} = 0$ *if* $s < m$.

(ii) $a_{n+m,n}^{(m)} = \prod_{i=0}^{m-1} a_{n+i+1,n+i}$.

Proof of (*i*) (by induction on s). If $s = 1$, then for all n, $a_{n+m,n}^{(1)} = a_{n+m,n} = 0$, if $1 < m$, since $[a_{jk}]$ is almost superdiagonal.

Now let $s > 1$ and assume that for any n, $a^{(s-1)}_{n+m,n} = 0$ whenever $s - 1 < m$. Fixing n, the definition of matrix multiplication gives for any m

$$\text{(iii)}\quad a^{(s)}_{n+m,n} = \sum_{i=1}^{\infty} a^{(s-1)}_{n+m,i}\, a_{i,n}.$$

Suppose $s < m$. Then $s - 1 < m - 1$ and by the induction hypothesis, $a^{(s-1)}_{n+m,i} = 0$ for $i \leqq n + 1$. But for $i > n + 1$, $a_{i,n} = 0$ since $[a_{jk}]$ is almost superdiagonal; hence all the terms under the summation sign in (iii) must be equal to 0, so $a^{(s)}_{n+m,n} = 0$, establishing (i).

Proof of (ii) (by induction on m). For the case $m = 1$,

$$a^{(1)}_{n+1,n} = a_{n+1,n} = \prod_{i=0}^{0} a_{n+i+1,n+i}.$$

Now let $m > 1$ and assume that for all n,

$$a^{(m-1)}_{n+m-1,n} = \prod_{i=0}^{m-2} a_{n+i+1,n+i}.$$

Fixing n, we have by the definition of matrix multiplication,

$$a^{(m)}_{n+m,n} = \sum_{i=1}^{\infty} a^{(m-1)}_{n+m,i} a_{i,n}.$$

For $i \leqq n$, $a^{(m-1)}_{n+m,i} = 0$ by (i), and for $i > n + 1$, $a_{i,n} = 0$ since $[a_{jk}]$ is almost superdiagonal. Therefore only the $(n + 1)$-st term in the summation is nonzero, and using this together with the induction hypothesis,

$$\begin{aligned} a^{(m)}_{n+m,n} &= a^{(m-1)}_{n+m,n+1} a_{n+1,n} = a^{(m-1)}_{n+1+m-1,n+1} a_{n+1,n} \\ &= \left(\prod_{i=0}^{m-2} a_{n+1+i+1,n+1+i}\right) a_{n+1,n} \\ &= \left(\prod_{i=1}^{m-1} a_{n+i+1,n+i}\right) a_{n+1,n} = \prod_{i=0}^{m-1} a_{n+i+1,n+i}. \end{aligned}$$

This establishes (ii) proving Theorem 4.2.

THEOREM 4.3: *Let T be a bounded linear operator on H which is represented by an almost superdiagonal matrix $[a_{jk}]$. Let*

$$p(z) = c_0 + c_1 z + \cdots + c_m z^m, \qquad c_m \neq 0, \quad m \geqq 1,$$

be a polynomial with standard complex coefficients such that $p(T)$ is compact. Then there exists an infinite positive integer ν such that $a_{\nu+1,\nu}$ is infinitesimal.

Proof: The operator $p(T)$ is represented by a matrix $[b_{jk}]$. For any $m, n \in Z$

$$b_{n+m,n} = c_1 a_{n+m,n} + c_2 a^{(2)}_{n+m,n} + \cdots + c_m a^{(m)}_{n+m,n}$$
$$= c_m a^{(m)}_{n+m,m} = c_m \prod_{i=0}^{m-1} a_{n+i+1,n+i}$$

by Theorem 4.2. The above equality holds also for infinite n since it can be expressed by a sentence of L. But it follows from Theorem 4.1 that $b_{n+m,n}$ is infinitesimal for infinite n. Since c_m is not infinitesimal one of the other factors in the preceding expression for $b_{n+m,n}$ must be infinitesimal, say $a_{n+j+1,n+j}$, $0 \leqq j < m$. Setting $\nu = n + j$ we obtain the theorem.

5. SUBSPACES IN NON-STANDARD HILBERT SPACE

Let $\{e_i\}$ be a basis for H and let ν be an infinite integer. Defining H_ν and P_ν as before, let E be an (internal) linear subspace of H_ν. Define $°E$, a subset of H, by: $x \in °E$ iff $x \in H$ and $\|x - x'\|$ is infinitesimal for some $x' \in E$. Let P_E denote the projection of $*H$ onto E. Then since P_E is the nearest point operator onto E, $\|x - x'\| \geqq \|x - P_E x\|$ for all x in H and $x' \in E$, and it follows that for x in H, $x \in °E$ iff $\|x - P_E x\|$ is infinitesimal.

THEOREM 5.1: *Given E as above, $°E$ is a closed linear subspace of H.*

Proof: Let $x, y \in °E$ and $\lambda \in C$. There exist elements x', y' in E such that $\|x - x'\|$ and $\|y - y'\|$ are infinitesimal. Then $x' + y' \in E$ and

$$\|(x + y) - (x' + y')\| \leqq \|x - x'\| + \|y - y'\|,$$

so that the left hand side of this inequality is also infinitesimal. Hence $x + y$ belongs to $°E$. Also, $\lambda x' \in E$ and

$$\|\lambda x - \lambda x'\| = |\lambda| \, \|x - x'\|,$$

is infinitesimal, so $\lambda x \in °E$. This shows that $°E$ is linear in the algebraic sense.

Now let $x_n \to x$ where the x_n are defined for standard $n \in Z$ and belong to $°E$. In order to prove that $°E$ is closed we have to show that x belongs to $°E$. For each $n \in Z$,

$$\|P_E x - x\| \leqq \|x - x_n\| + \|x_n - P_E x_n\|.$$

For a given standard $\varepsilon > 0$, choose $n \in Z$ such that $\|x - x_n\| < \varepsilon/2$. Since for each $n \in Z$, $x_n \in °E$ and consequently $\|x_n - P_E x_n\|$ is infinitesimal, it follows that $\|P_E x - x\| \leqq \varepsilon$. Since ε was an arbitrary standard positive number, $\|P_E x - x\|$ is infinitesimal, hence $x \in °E$ and the proof is complete.

Let T be a bounded linear operator on H and let ν be an infinite integer. We define the operator T_ν on H_ν to be the restriction of $P_\nu{}^*TP_\nu$ to H_ν. Then

$$\|T_\nu\| \leqq \|P_\nu\|^2 \, \|{}^*T\| \leqq \|{}^*T\| = \|T\|,$$

so that T_ν has finite norm.

THEOREM 5.2: *Let E be an internal linear subspace of H_ν which is invariant for T_ν, i.e., $T_\nu E \subseteq E$. Then $°E$ is invariant for T, $T\,°T \subseteq °E$.*

Proof: Choose any $x \in °E$. Then $P_E x \sim x$ and since ${}^*T$ has finite norm, ${}^*TP_E x \sim {}^*Tx$. Since ${}^*Tx$ is standard, ${}^*TP_E x$ is near-standard and we may apply Theorem 3.7 to conclude that $P_\nu {}^*TP_E x \sim {}^*TP_E x$. Putting the last two relations together, $T_\nu P_E x = P_\nu {}^*TP_E x \sim {}^*Tx = Tx$ where $T_\nu P_E x$ is in E since E is invariant for T_ν. Tx is thus infinitely close to an element of E, so $Tx \in °E$. This shows that $°E$ is invariant for T which proves the theorem.

The number of dimensions of H_ν as defined within the language L is ν, $d(H_\nu) = \nu$. In this sense H_ν is "finite-dimensional." Similarly, with every (internal) linear subspace E of H_ν there is associated an integer $d(E) \in {}^*Z$ (or $d(E) = 0$) which may be finite or infinite, and which has the properties of a dimension to the extent to which these can be expressed as sentences of L.

THEOREM 5.3: *Let E, E_1 be linear subspaces of H_ν such that $E \subset E_1$ and $d(E_1) = d(E) + 1$. Then ${}^\circ E \subseteq {}^\circ E_1$ and any two points of ${}^\circ E_1$ are linearly dependent modulo ${}^\circ E$.*

Proof: Since $E \subset E_1$, it is trivial that ${}^\circ E \subseteq {}^\circ E_1$. Now let x, $y \in {}^\circ E_1$. There exist x', y' in E_1 such that $x \sim x'$ and $y \sim y'$. Since the dimension of E_1 exceeds that of E only by 1 there is a representation

(5.1) $x' = \lambda y' + z$ or vice versa, where $\lambda \in {}^*C$ and $z \in E$.

If λ is finite, then it possesses a standard part ${}^\circ\lambda$ and $\lambda y'$ is infinitely close to the standard point ${}^\circ\lambda y$. Therefore $z = z' - \lambda y'$, being the difference of two near-standard points, must itself possess a standard part ${}^\circ z$, ${}^\circ z \sim z$. Now $x' \sim x$, $\lambda y' \sim {}^\circ\lambda y$, and $z \sim {}^\circ z$, so substituting this in (5.1) we obtain $x \sim {}^\circ\lambda y + {}^\circ z$ where ${}^\circ z \in {}^\circ E$ since $z \in E$. But two standard numbers cannot be infinitely close unless they are equal, so $x = {}^\circ\lambda y + {}^\circ z$ and therefore x and y are linearly dependent modulo ${}^\circ E$.

If λ is infinite then we may rewrite (5.1) as, $y' = (1/\lambda)x' + (-1/\lambda)z$. Now $1/\lambda$ is infinitesimal, *a fortiori* finite, and $(-1/\lambda)z \in E$ so we have precisely the same case as was considered in the previous paragraph with y and x interchanged. Therefore we again conclude that y and x are linearly dependent modulo ${}^\circ E$, which completes the proof of the theorem.

6. MAIN THEOREM

We are now ready to prove the main result.

THEOREM 6.1: *Let T be a bounded linear operator on an infinite-dimensional Hilbert space H over the complex numbers and let $p(z) \neq 0$ be a polynomial with complex coefficients such that $p(T)$ is compact. Then T leaves invariant at least one closed linear subspace of H other than H or $\{0\}$.*

Proof: The method of proof, like that of [1] is based on the fact that in a *finite-dimensional* space, of dimension m say, any linear operator possesses a chain of invariant subspaces

$$E_0 \subset E_1 \subset E_2 \subset \cdots \subset E_m, \tag{6.1}$$

where $d(E) = j$, $0 \leqq j \leqq m$, so $E_0 = \{0\}$ (cf. [5]).

Let $x \in H$, $\|x\| = 1$. The proof is trivial unless the set

$$A = \{x, Tx, T^2x, \cdots\},$$

is linearly independent algebraically and generates the entire space H. Otherwise A generates a proper invariant subspace for T. Assuming from now on that this is the case we replace A by an equivalent orthonormal set $B = \{x = e_1, e_2, e_3, \cdots\}$ by the Gram-Schmidt method. Thus $\{e_i\}$ is an orthonormal basis for H and for any $n \in Z$,

$$\text{span } \{e_1, \cdots, e_n\} = \text{span } \{x, Tx, T^2x, \cdots, T^{n-1}x\}.$$

For any integer $k \in Z$,

$$e_k \in sp\{e_1, e_2, \cdots, e_k\} = sp\{x, Tx, \cdots, T^{k-1}x\},$$

so that

$$Te_k \in sp\{Tx, T^2x, \cdots, T^kx\} \subseteq sp\{e_1, e_2, \cdots, e_{k+1}\}.$$

Thus $Te_k = \sum_{j=1}^{k+1} \beta_{jk}e_j$ for some $\beta_{jk} \in C$, $0 < j \leqq k + 1$. After choosing such β_{jk} for each $k \in Z$, we define the matrix $[a_{jk}]$ by (i) $a_{jk} = \beta_{jk}$ if $j \leqq k + 1$, and (ii) $a_{jk} = 0$ if $j > k + 1$. Then $[a_{jk}]$ is an almost superdiagonal matrix which represents T relative to the basis $\{e_i\}$.

Now suppose that T is a bounded linear operator on H and that

$$p(\lambda) = c_0 + c_1\lambda + c_2\lambda^2 + \cdots + c_n\lambda^n, \qquad c_n \neq 0,$$

is a complex polynomial such that $p(T)$ is compact. Using Theorem 4.3 let ν be an infinite integer such that $a_{\nu+1,\nu}$ is infinitesimal. For brevity we denote by P the projection P_ν of $^*H$ onto H_ν and by T' operator P^*TP. We next show by induction on finite integers n that

$$(^*T)^n\xi \sim (T')^n\xi, \tag{6.2}$$

4. APPLICATIONS TO TWO-CARDINAL THEOREMS

Let L have a 1-place relation symbol U, which shall be distinguished from the other symbols of L. Let T be a set of sentences of L. We say that T *admits the pair* (α, β) iff T has a model $\mathfrak{A} = \langle A, V, \cdots \rangle$ where V is the interpretation of U in $\mathfrak{A}$ and $|A| = \alpha$, $|V| = \beta$. A two-cardinal theorem is a statement of the following sort: If T admits (α, β), then T admits (γ, δ), where appropriate conditions are put upon α, β, γ, δ. Many such two-cardinal theorems are known; some are very easy, others are extremely difficult. There are still several problems in this area. We'll ask the interested reader to peruse elsewhere for surveys of all known two-cardinal theorems (a good place to start is again Chang-Keisler 1973 [7] and, also, Vaught 1965 [20]). Our purpose in this section is to give two essential applications of saturated and special models to two-cardinal theorems. Let us first mention a theorem of Vaught (see Morley-Vaught 1962 [15]):

THEOREM 4.1: *If T admits (α^+, α), then it admits (ω_1, ω).*

This result does not use saturated or special models, but it does use the closely related notion of countable homogeneous models. The basic idea in the proof is to build a tower of models. This idea is also used in the next theorem, an essential application of saturated models. (Essential application means no other proof is known.)

THEOREM 4.2: (Chang 1965 [4].) (GCH). *If T admits (α^+, α) and δ is regular, then T admits (δ^+, δ).*

Proof: By Theorem 4.1, we may assume that δ is a regular cardinal greater than ω. Let $\mathfrak{A} = \langle A, V, \cdots \rangle$ be a model of T such that $|A| = \alpha^+$ and $|V| = \alpha$. Let R be a 2-place relation over V such that R indexes all the finite subsets of V. We may express this in the expanded language $L \cup \{P\}$, P a new 2-place relation symbol, by: for all $u_1, \cdots, u_n \in V$, there exists a $u \in V$ such that

(1) $(\mathfrak{A}, R) \models \forall t(U(t) \rightarrow (P(ut) \leftrightarrow t \equiv u_1 \vee \cdots \vee t \equiv u_n))$.

By the downward Löwenheim-Skolem theorem, there is an elementary submodel $\langle B, V, R, \cdots \rangle$ of $(\mathfrak{A}, R)$ such that $V \subset B$

and $|B| = \alpha$. Let Q be a new 1-place relation symbol, and consider the model

$$(\mathfrak{A}, B, R) = \langle A, B, V, R, \cdots \rangle,$$

for $L \cup \{P, Q\}$, the interpretation of Q in $\mathfrak{A}$ is the set B. Let Σ be the set of all sentences of $L \cup \{P, Q\}$ holding in $(\mathfrak{A}, B, R)$. Note that, by (1), Σ contains all closures of formulas of the form

$$(2)\quad U(z_1) \wedge \cdots \wedge U(z_n) \wedge \varphi(z_1 x_1 \cdots x_m) \wedge \cdots \wedge \varphi(z_n x_1 \cdots x_m) \rightarrow (\exists y)(P(yz_1) \wedge \cdots \wedge P(yz_n) \wedge (\forall t)(P(yt) \rightarrow \varphi(tx_1 \cdots x_m))),$$

where $\varphi(zx_1 \cdots x_m)$ is an arbitrary formula of $L \cup \{P, Q\}$. By Theorem 1.5 (i), Σ has a saturated model of power δ, say $\langle A_0, B_0, V_0, R_0, \cdots \rangle$. It is clear that $|A_0| = |B_0| = |V_0| = \delta$. Since B gives rise to a proper elementary submodel of $(\mathfrak{A}, R)$, it follows that B_0 gives rise to the model $\mathfrak{B}_0 = \langle B_0, V_0, R_0, \cdots \rangle$ which is a proper elementary submodel of the model

$$\mathfrak{A}_0 = \langle A_0, V_0, R_0, \cdots \rangle.$$

Now, both $\mathfrak{B}_0$ and $\mathfrak{A}_0$ are equivalent saturated models of power δ, whence by Theorem 1.8, $\mathfrak{A}_0 \cong \mathfrak{B}_0$.

Let us now fix the model $\mathfrak{A}_0 = \langle A_0, V_0, R_0, \cdots \rangle$. An arbitrary model $\mathfrak{B} = \langle B, W, S, \cdots \rangle$ of power δ is called *W-saturated* iff every type $\Sigma(x)$ over every expansion $(\mathfrak{B}, b_\xi)_{\xi<\eta<\delta}$ which contains the formula $U(x)$ is realized in $(\mathfrak{B}, b_\xi)_{\xi<\eta<\delta}$. The models $\mathfrak{A}_0$, $\mathfrak{B}_0$ are V_0-saturated. It is clear, by using the back and forth argument, that

(3) if the model $\mathfrak{B} = \langle B, W, S, \cdots \rangle$ is equivalent to $\mathfrak{B}_0$, is of power δ, and is W-saturated, then $\mathfrak{B}$ can be elementarily embedded in $\mathfrak{B}_0$ with W mapped *onto* V. Whence, since $\mathfrak{A}_0$ is a proper elementary extension of $\mathfrak{B}_0$, $\mathfrak{B}$ has a proper elementary extension $\overline{\mathfrak{B}}$ of power δ with the same W and $\overline{\mathfrak{B}}$ is still W-saturated.

We now seek to construct a strictly increasing elementary tower of models $\mathfrak{A}_\xi$, $\xi < \delta^+$, such that each $\mathfrak{A}_\xi = \langle A_\xi, V_0, R_0, \cdots \rangle$ is of power δ and is V_0-saturated. $\mathfrak{A}_0$ is already defined. By (3), there is obviously no trouble going from $\mathfrak{A}_\xi$ to $\mathfrak{A}_{\xi+1}$. So suppose that λ is a limit ordinal, $\lambda < \delta^+$ and the models $\mathfrak{A}_\xi$, $\xi < \lambda$, are found. Let $\mathfrak{A}_\lambda = \bigcup_{\xi<\lambda} \mathfrak{A}_\xi$. Clearly $\mathfrak{A}_\lambda = \langle \bigcup_{\xi<\lambda} A_\xi, V_0, R_0, \cdots \rangle$ is a model of power δ, and is a proper elementary extension of each $\mathfrak{A}_\xi$, $\xi < \lambda$. We next show that

(4) $\mathfrak{A}_\lambda$ is V_0-saturated.

This is the key to the whole proof. Let $\Sigma(x)$ be a type over the expansion $\overline{\mathfrak{A}}_\lambda = (\mathfrak{A}_\lambda, a_\xi)_{\xi<\eta<\delta}$. Whence, $\Sigma(x)$ is finitely satisfiable in $\overline{\mathfrak{A}}_\lambda$ by elements of V_0. Let s be any finite subset of Σ, and let $\sigma_s(x)$ be the conjunction of the formula in s together with the formula $U(x)$. Since $\sigma_s(x)$ is realized in $\overline{\mathfrak{A}}_\lambda$ by an element $u_s \in V_0$, and since $\sigma_s(x)$ contains only a finite number of constants, say $a_{\xi_1}, \cdots, a_{\xi_n}$, let v_s be an ordinal less than λ such that $a_{\xi_1}, \cdots, a_{\xi_n} \in A_{v_s}$. Whence

$$\mathfrak{A}_{v_s} \models \sigma_s[u_s a_{\xi_1} \cdots a_{\xi_n}],$$

where $\sigma_s(y x_1 \cdots x_n)$ is a formula of $L \cup \{P\}$. If $t \supset s$ is another finite subset of Σ, we see that the corresponding u_t will also belong to $\mathfrak{A}_{v_s}$ and so

$$\mathfrak{A}_{v_s} \models \sigma_s[u_t a_{\xi_1} \cdots a_{\xi_n}].$$

For each fixed finite subset s of Σ, let

$$\Gamma_s(y) = \{P(y, u_t)\colon s \subset t,\ t \text{ a finite subset of } \Sigma\}$$
$$\cup \{(\forall z)(P(yz) \to \sigma_s(z a_{\xi_1} \cdots a_{\xi_n}))\} \cup \{U(y)\}.$$

Since $|\Sigma| < \delta$, $|\Gamma_s| < \delta$. By (2), Γ_s is finitely satisfiable in (an expansion of) $\mathfrak{A}_{v_s}$ by an element of V_0. Since $\mathfrak{A}_{v_s}$ is V_0-saturated, Γ_s is realized in $\mathfrak{A}_{v_s}$ by an element w_s in V_0. The element w_s has the properties

$$\mathfrak{A}_{v_s} \models (\forall z)(P(w_s z) \to \sigma_s(z a_{\xi_1} \cdots a_{\xi_n})).$$

and

$$\mathfrak{A}_{v_s} \models P(w_s, u_t) \quad \text{for all} \quad t \supset s.$$

Let $\Delta(z) = \{U(z)\} \cup \{P(w_s z)\colon s \text{ finite subset of } \Sigma\}$. Then $|\Delta| < \delta$ and Δ is a type over (an expansion of) $\mathfrak{A}_0$. Since $\mathfrak{A}_0$ is V_0-saturated, Δ is realized in $\mathfrak{A}_0$ by an element $u \in V_0$. It is easy to see that this element u must realize Σ in $\overline{\mathfrak{A}}_\lambda$. So (4) is proved and the induction is complete. The model $\bigcup_{\xi<\delta^+} \mathfrak{A}_\xi$ is clearly a model of T with the desired property (we may drop the superfluous relation R_0). $\dashv$

Not only is the use of saturated models in Theorem 4.2 essential, it is also known that there can be no proof of Theorem 4.2 without the GCH. (See Chang-Keisler for fuller references.) The intriguing problem of whether Theorem 4.2 can be improved so that the requirement that δ be regular is removed is still an open problem. The best result we know of in this direction is the following theorem of Jensen which assumes the axiom of constructibility:

THEOREM 4.3: (Jensen 1970 [10].) ($V = L$) *If T admits (α^+, α), then T admits every (δ^+, δ).*

The proof of this theorem is rather long and involved. Jensen first proved that under $V = L$ a certain combinatorial principle holds for each infinite cardinal α, then he deduces the theorem 4.3 from the combinatorial principle. It is this second implication that requires another essential use of saturated and special models. Since we do not wish to be involved in a great deal of set theory, we shall simply conclude this survey with a sketch of a proof of the more straightforward second implication.

Let α be an infinite cardinal. Let $\square_\alpha$ (Jensen's notation) denote the following combinatorial property of α: There exists a family $\{C_\xi : \xi < \alpha^+$ and ξ is a limit ordinal$\}$ of subsets of α^+ such that

(i) each C_ξ is closed and unbounded in ξ,
(ii) if $cf(\xi) < \alpha$, then $|C_\xi| < \alpha$,
(iii) if λ is a limit point of C_ξ, then $C_\xi \cap \lambda = C_\lambda$.

Jensen actually proved that:

(1) $V = L$ implies $\square_\alpha$ for every infinite cardinal α, and
(2) $\square_\alpha$ and $\alpha = \sum_{\beta<\alpha} 2^\beta$ imply that any set of sentences admitting (γ^+, γ) admits (α^+, α).

Even his proof of (2) involves a fair amount of argument about collapsing cardinals so that the GCH holds below α. We shall simply prove the essential point of (2), namely:

(3) Let $\alpha = \aleph_\omega$ and assume that $\aleph_{n+1} = 2^{\aleph_n}$ and that $\square_\alpha$. Then any set of sentences admitting (γ^+, γ) admits (α^+, α).

Proof of (3): Let T be any set of sentences which admits (γ^+, γ).

We may as well assume that the 2-place relation R indexing all finite subsets of V is already present in L and that T already contains all the sentences in (2) of Theorem 4.2. In Theorem 4.2 we constructed a strictly increasing elementary tower of models $\mathfrak{A}_\xi = \langle A_\xi, V, \cdots \rangle$, $\xi < \delta^+$, so that each $\mathfrak{A}_\xi$ is of power δ and is V-saturated, and for limit ordinals λ, $\mathfrak{A}_\lambda = \bigcup_{\xi<\lambda} \mathfrak{A}_\xi$. Instead of starting with a saturated model of power δ, the obvious thing to do here is to start with a special model $\mathfrak{A} = \bigcup_n \mathfrak{A}_n$ of power $\alpha = \aleph_\omega$, and each model $\mathfrak{A}_n$ ($n \geqq 1$) is a saturated model of power $\aleph_n$ (GCH). Let the model $\mathfrak{A}$ be fixed, as well as each $\mathfrak{A}_n = \langle A_n, V_n, \cdots \rangle$. We say that a model $\mathfrak{B}$ of power $\aleph_\omega$ is of the *correct form* iff $\mathfrak{B} \equiv \mathfrak{A}$, $\mathfrak{B} = \bigcup_n \mathfrak{B}_n$, each $\mathfrak{B}_n$, ($1 \leqq n$) is of power $\aleph_n$, $\mathfrak{B}_n = \langle B_n, V_n, \cdots \rangle$, and each $\mathfrak{B}_n$ is V_n-saturated. It is clear, by looking at the proof of Theorem 4.2, that if $\mathfrak{B}$ is of the correct form, then $\mathfrak{B}$ can be mapped properly into an elementary submodel of $\mathfrak{A}$ with each $\mathfrak{B}_n$ going into $\mathfrak{A}_n$ and with V_n going onto V_n. Whence, $\mathfrak{B}$ has a proper elementary extension of the correct form and with the same set $V = \bigcup_n V_n$.

The plan of the proof, therefore, is to construct a strictly increasing elementary tower of models $\mathfrak{A}^\xi$, $\xi < \aleph_{\omega+1}$, starting with $\mathfrak{A}^0 = \mathfrak{A}$ such that each $\mathfrak{A}^\xi$ is of the correct form, $\mathfrak{A}^\xi = \bigcup_n \mathfrak{A}^\xi_n$. This amounts to defining the double array $\mathfrak{A}^\xi_n$, $0 \leqq \xi < \aleph_{\omega+1}$, $1 \leqq n < \omega$ by transfinite induction on ξ.

By the assumption $\square_\alpha$, we know that there are sets C_ξ, ξ limit ordinal less than α^+ satisfying (i), (ii), (iii) above. Since $cf(\xi) < \alpha$ for any $\xi < \alpha^+$, (ii) always gives us $|C_\xi| < \alpha$ and hence $|C_\xi| = \aleph_n$ for some $n < \omega$. Given two models $\mathfrak{A}^\xi$, $\mathfrak{A}^\eta$ both of the correct form, we say that *eventually* $\mathfrak{A}^\xi = \mathfrak{A}^\eta$ iff there is an m such that for all $n \geqq m$, $\mathfrak{A}^\xi_n = \mathfrak{A}^\eta_n$; similarly we say that *eventually* $\mathfrak{A}^\xi \prec \mathfrak{A}^\eta$ iff there is an m such that for all $n \geqq m$, $\mathfrak{A}^\xi_n \prec \mathfrak{A}^\eta_n$.

We shall define by transfinite induction the sequence of models $\mathfrak{A}^\xi = \bigcup_n \mathfrak{A}^\xi_n$, $\xi < \alpha^+$, such that:

(a) each $\mathfrak{A}^\xi$ is of the correct form so that each $\mathfrak{A}^\xi_n$, $1 \leqq n$, is of power $\aleph_n$ and is V_n-saturated;
(b) $\mathfrak{A}^\xi \prec \mathfrak{A}^\eta$ if $\xi \leqq \eta$, and properly if $\xi < \eta$;
(c) if $\xi \leqq \eta$, then eventually $\mathfrak{A}^\xi \prec \mathfrak{A}^\eta$.

In addition to (a), (b), (c), there is another condition that the models $\mathfrak{A}^\xi$ must satisfy for all limit ordinals ξ. This condition is complicated enough so that we shall only state it when we come to construct $\mathfrak{A}^\xi$ for a limit ξ.

The model $\mathfrak{A}^0$ is already defined. By an earlier remark, it is easy to go from $\mathfrak{A}^\xi$ to $\mathfrak{A}^{\xi+1}$ so that (a), (b), (c) are preserved. Suppose that $\mathfrak{A}^\xi$ are constructed for all $\xi < \lambda < \alpha^+$ and λ is a limit ordinal. To construct $\mathfrak{A}^\lambda$, we shall first construct a new array of models $\mathfrak{A}^{\lambda,\xi} = \bigcup_n \mathfrak{A}_n^{\lambda,\xi}$, $\xi \in C_\lambda$, so that the following hold:

(d) each $\mathfrak{A}^{\lambda,\xi}$ is of the correct form so that each $\mathfrak{A}_n^{\lambda,\xi}$, $1 \leqq n$, is of power $\aleph_n$ and is V_n-saturated;
(e) for each $\xi \in C_\lambda$, eventually $\mathfrak{A}^\xi = \mathfrak{A}^{\lambda,\xi}$;
(f) let $|C_\lambda| = \aleph_r$, then for all $1 \leqq n \leqq r$, and all $\eta \in C_\lambda$, $\mathfrak{A}_n^{\lambda,\eta} = \mathfrak{A}_n^{\xi_0}$ where ξ_0 is the first element of C_λ;
(g) if $m \leqq n$, $\xi \leqq \eta$, and $\xi, \eta \in C_\lambda$, then $\mathfrak{A}_m^{\lambda,\xi} \prec \mathfrak{A}_n^{\lambda,\eta}$;
(h) if ρ is a limit point of C_λ, then for each n,

$$\mathfrak{A}_n^{\lambda,\rho} = \bigcup_{\xi \in C_\rho} \mathfrak{A}_n^{\lambda,\xi}.$$

These conditions look horrible, but actually they are quite understandable. The point is that we cannot simply define each $\mathfrak{A}_n^\lambda = \bigcup_{\xi<\lambda} \mathfrak{A}_n^\xi$, because $|\lambda|$ may be so large that $|\mathfrak{A}_n^\lambda|$ may now exceed $\aleph_n$. So naturally the next best thing to do is to use the set C_λ by defining (assume $|C_\lambda| = \aleph_r$)

$$\mathfrak{A}_n^\lambda = \mathfrak{A}_n^{\xi_0} \quad \text{for} \quad n \leqq r,$$

$$\mathfrak{A}_n^\lambda = \bigcup_{\xi \in C_\lambda} \mathfrak{A}_n^\xi \quad \text{for} \quad n > r.$$

This second definition will make sure that $|\mathfrak{A}_n^\lambda| = \aleph_n$, however, the first line will destroy the nice property that for a fixed n the $\mathfrak{A}_n^\xi$'s are increasing as ξ increases, and so the second line is no longer a union of a chain. Whence, the need for the roundabout procedure we are describing in (d)–(h) and in what follows. The models $\mathfrak{A}^{\lambda,\xi}$ satisfying (d)–(h) will be defined by induction on $\xi \in C_\lambda$.

For the first ξ_0 in C_λ, we let $\mathfrak{A}_n^{\lambda,\xi_0} = \mathfrak{A}_n^{\xi_0}$. Let $|C_\lambda| = \aleph_r$, $r < \omega$. For any $n \leqq r$ and any $\xi \in C_\lambda$, define

$$\mathfrak{A}_n^{\lambda,\xi} = \mathfrak{A}_n^{\xi_0}. \tag{4}$$

Suppose $\mathfrak{A}^{\lambda,\xi}$ is defined. Let η be the next ordinal in C_λ. By (4), $\mathfrak{A}_n^{\lambda,\eta}$ are already defined for $n \leqq r$. So let $n > r$, and define

(5) $$\mathfrak{A}_n^{\lambda,\eta} = \mathfrak{A}_n^\eta \quad \text{if for all} \quad m \geqq n, \ \mathfrak{A}_m^{\lambda,\xi} \prec \mathfrak{A}_m^\eta,$$
$$= \mathfrak{A}_n^{\lambda,\xi} \text{ otherwise.}$$

If ρ is a limit point of C_λ, then for all n, define

(6) $$\mathfrak{A}_n^{\lambda,\rho} = \bigcup_{\xi \in C_\rho} \mathfrak{A}_n^{\lambda,\xi}.$$

We finally define

$$\mathfrak{A}_n^\lambda = \bigcup_{\xi \in C_\lambda} \mathfrak{A}_n^{\lambda,\xi} \quad \text{for all} \quad n.$$

Notice that the definition of the $\mathfrak{A}_n^{\lambda,\xi}$ given in (4), (5), (6) in no way depends on the verification of (d)–(h) for the models $\mathfrak{A}^{\lambda,\xi}$. So this is the construction of $\mathfrak{A}^\lambda$ for limit ordinals λ that we referred to earlier. We shall now verify that $\mathfrak{A}^\lambda$ satisfies (a), (b), (c), by proving that (d)–(h) hold by induction on $\xi \in C_\lambda$. Here we assume of course that for any earlier limit ordinal ρ, the model $\mathfrak{A}^\rho$ satisfies (a)–(c) and the models $\mathfrak{A}^{\rho,\xi}$, $\xi \in C_\rho$, satisfy (d)–(h).

By the proof of Theorem 4.2 (roughly, the union of an elementary chain of V_n-saturated models is V_n-saturated) it is pretty clear that the definitions (4), (5), (6) give easily the conditions (d), (f), (g), (h). The hard one to verify is (e), which we shall now do by induction on $\xi \in C_\lambda$. Clearly (e) is true for ξ_0. If (e) is true for $\xi \in C_\lambda$, then by (5) and (c), (e) is true for the next ordinal in C_λ. So let ρ be a limit point in C_λ and assume (e) holds for all $\xi \in \rho \cap C_\lambda = C_\rho$. Since C_λ is closed in λ, ρ is a limit ordinal, so by our construction,

$$\mathfrak{A}_n^\rho = \bigcup_{\xi \in C_\rho} \mathfrak{A}_n^{\rho,\xi}.$$

Let $|C_\rho| = \aleph_q$. We have $q \leqq r$. We show that

(7) for all $\xi \in C_\rho$, $\mathfrak{A}_n^{\lambda,\xi} = \mathfrak{A}_n^{\rho,\xi}$ for all $n > r$.

This requires a second induction on all $\xi \in C_\rho$. Again (7) holds for ξ_0, the first element of C_ρ. Suppose (7) holds for some $\xi \in C_\rho$ and let η be the next ordinal in C_ρ (as well as in C_λ). By (5), we have

$$\text{for } n \leqq r, \quad \mathfrak{A}_n^{\lambda,\eta} = \mathfrak{A}_n^{\xi_0};$$

$$\text{for } n > r, \quad \begin{cases} \mathfrak{A}_n^{\lambda,\eta} = \mathfrak{A}_n^{\eta} & \text{if for all } m \geqq n,\ \mathfrak{A}_m^{\lambda,\xi} \prec \mathfrak{A}_m^{\eta}, \\ \quad\ \ = \mathfrak{A}_n^{\lambda,\xi} & \text{otherwise;} \end{cases}$$

$$\text{for } n \leqq q, \quad \mathfrak{A}_n^{\rho,\eta} = \mathfrak{A}_n^{\xi_0};$$

$$\text{for } n > q, \quad \begin{cases} \mathfrak{A}_n^{\rho,\eta} = \mathfrak{A}_n^{\eta} & \text{if for all } m \geqq n,\ \mathfrak{A}_m^{\rho,\xi} \prec \mathfrak{A}_m^{\eta}, \\ \quad\ \ = \mathfrak{A}_n^{\rho,\xi} & \text{otherwise.} \end{cases}$$

Since (7) already holds for ξ, the above few lines show that (7) must also hold for η. Suppose σ is a limit point of C_ρ. By (6), we get

$$\mathfrak{A}_n^{\rho,\sigma} = \bigcup_{\xi \in C_\sigma} \mathfrak{A}_n^{\rho,\xi} \quad \text{and} \quad \mathfrak{A}_n^{\lambda,\sigma} = \bigcup_{\xi \in C_\sigma} \mathfrak{A}_n^{\lambda,\xi},$$

which shows that (7) holds for σ. So (7) is proved. Now (e) holds for ρ. So (e) holds. From (e) and the definition of $\mathfrak{A}^\lambda$, we easily get (a)–(c).

Finally, the model $\bigcup_{\xi<\alpha^+} \mathfrak{A}^\xi$ is a (α^+, α) model of T. This concludes the proof of (3). ⊣

Added in proof, April 16, 1973. Recently Jensen has improved Theorem 4.3 by showing that the Gap n Conjecture holds in L. The degree to which he makes essential use of saturated and special models is no more than the use of such models in Theorem 4.3; however, the complexities of the combinatorial arguments increase exponentially. (Gap n Conjecture says that if T admits $(\aleph_n(\alpha), \alpha)$, then T admits $(\aleph_n(\beta), \beta)$.)

REFERENCES

1. Beth, E. W., "On Padoa's method in the theory of definition," *Nederl. Akad. Wetensch. Proc. Ser. A*, **56** (Indag. Math., 15), (1953), 330–339.

2. Chang, C. C., "On unions of chains of models," *Proc. Amer. Math. Soc.*, **10** (1959), 120–127.

3. ———, "Some new results in definability," *Bull. Amer. Math. Soc.*, **70** (1964), 808–813.

4. ———, "A note on the two-cardinal problem," *Proc. Amer. Math. Soc.*, **16** (1965), 1148–1155.

5. ———, "A generalization of the Craig interpolation theorem," *Amer. Math. Soc. Notices*, **15** (1968), 934.

6. ———, "Two interpolation theorems," *Symposia Mathematica V*, Academic Press, 1971, 1–19.

7. Chang, C. C., and H. J. Keisler. *Model Theory*. To be published by North Holland, 1973.

8. Craig, W., "Three uses of the Herbrand-Gentzen theorem in relating model theory and proof theory," *J. Symb. Logic*, **22** (1957), 269–285.

9. Kueker, D. W., "Generalized interpolation and definability," *Ann. Math. Logic*, **1** (1970), 423–468.

10. Jensen, R., "A note on the two-cardinal conjecture," Mimeographed notes.

11. Łoś, J., "On the extending of models, I," *Fund. Math.*, **42** (1955), 38–54.

12. Łoś, J., and R. Suszko, "On the extending of models, IV; Infinite sums of models," *Fund. Math.*, **44** (1957), 52–60.

13. Lyndon, R. C., "Properties preserved under homomorphisms," *Pacific J. Math.*, **9** (1959), 143–154.

14. Makkai, M., "On a generalization of a theorem of E. W. Beth," *Acta Math. Acad. Sci. Hungar.*, **15** (1964), 227–235.

15. Morley, M., and R. Vaught, "Homogeneous universal models," *Math. Scand.*, **11** (1962), 37–57.

16. Reyes, G. E., "Local definability theory," *Ann. Math. Logic*, **1** (1970), 95–137.

17. Robinson, A., "A result on consistency and its application to the theory of definition," *Nederl. Akad. Wetensch. Proc. Ser. A*, **59** (Indag. Math., **18**), (1956), 47–58. "On a problem of L. Henkin," *J. Symb. Logic*, **21** (1956), 33–35.

18. Shelah, S., "Remark to 'Local definability theory' of Reyes," *Ann. Math. Logic*, **2** (1971), 441–447.

19. Tarski, A., "Contributions to the theory of models. I, II," *Nederl. Akad. Wetensch. Proc. Ser. A*, **58** (Indag. Math., **16**), (1954), 572–588.

20. Vaught, R., "The Löwenheim-Skolem theorem," *Logic, Math. and Phil. of Sci. Proc. 1964 Int'l Congr.*, Y. Bar-Hillel, ed., Amsterdam, 1965, 81–89.

FORCING AND THE OMITTING TYPES THEOREM

H. Jerome Keisler

INTRODUCTION

Many results in model theory and set theory are proved by constructing a model from a set of formulas with constant symbols. This method, sometimes called the method of diagrams, was introduced by Henkin [13] and A. Robinson [29]. The Omitting Types Theorem was proved using diagrams by Henkin [14] and Orey [27]. It has a number of applications in model theory. The forcing construction in set theory, introduced by Cohen [5], also uses diagrams and has led to the solution of many classical problems by means of consistency results. A. Robinson [30] developed an analogous theory of forcing in model theory, and Barwise [3] extended Robinson's theory to infinitary logic and used it to give a new proof of the Omitting Types Theorem.

In this paper we shall give a general treatment which includes both the Omitting Types Theorem and Cohen's forcing. We shall emphasize the common ground of the two methods. We present a variety of applications, some selected from the extensive literature, and some new. Our treatment of forcing is actually a reformulation of the Boolean valued models of Rasiowa-Sikorski [28] in terms of ordinary models. The fact that Cohen forcing can be obtained using Boolean valued models was discovered by D. Scott and R. Solovay [32].

The basic result is the Generic Model Theorem (essentially due to Rasiowa-Sikorski [28]) in section 1. In section 2 we derive the Omitting Types Theorem and give some typical applications of it to model theory. Section 3 contains applications to group theory, and section 4 applications to second order number theory and Archimedean ordered fields. Finally, in section 5 we discuss Cohen forcing and give examples of the construction of models of set theory both by Cohen forcing and by the Omitting Types Theorem.

We are greatly indebted to Jon Barwise for many valuable discussions during the preparation of this article. This work was supported in part by National Science Foundation contract No. GP-27633.

1. GENERIC MODELS

Throughout this paper we let L be a countable first order predicate logic with equality. L has countably many variables $v_0, v_1, \cdots$, the connectives $\vee$ and $\neg$, the quantifier $\exists x$, the equality symbol $=$, and at most countably many relation, function, and constant symbols. The conjunction $\wedge$ and universal quantifier $\forall$ are regarded as abbreviations,

$$\varphi \wedge \psi \quad \text{for} \quad \neg(\neg\varphi \vee \neg\psi),$$

$$\forall x\varphi \quad \text{for} \quad \neg\exists x\neg\varphi.$$

Sentences are formulas without free variables, and theories are sets of sentences. Given a model M and theory T for L, $M \models T$ means that M is a model of T. T is *consistent* iff T has a model. $T \models \varphi(x)$ means that every model of T is a model of $\forall x\varphi(x)$.

The infinitary logic $L_{\omega_1\omega}$ is built from L by allowing the infinite disjunction $\bigvee\Phi$, or $\bigvee_{\varphi\in\Phi}\varphi$, of any countable set Φ of formulas. The infinite conjunction is again an abbreviation,

$$\bigwedge\Phi \quad \text{for} \quad \neg \bigvee_{\varphi\in\Phi} \neg\varphi.$$

The language $L_{\omega_1\omega}$ has uncountably many formulas. The proper setting for this paper is a well behaved countable part of the language $L_{\omega_1\omega}$. Let us explain what is meant by "well behaved." The set sub (φ) of subformulas of φ is defined recursively as follows.

If φ is atomic, sub $(\varphi) = \{\varphi\}$,

$$\text{sub}\,(\varphi \vee \psi) = \text{sub}\,(\varphi) \cup \text{sub}\,(\psi) \cup \{\varphi \vee \psi\},$$

$$\text{sub}\,(\neg\varphi) = \text{sub}\,(\varphi) \cup \{\neg\varphi\},$$

$$\text{sub}\,(\exists x\varphi) = \text{sub}\,(\varphi) \cup \{\exists x\varphi\},$$

$$\text{sub}\,(\bigvee\Phi) = \bigcup_{\varphi\in\Phi} \text{sub}\,(\varphi) \cup \{\bigvee\Phi\}.$$

It is easily seen that sub (φ) is at most countable. Given a formula $\varphi(x)$ and a term τ, $\varphi(\tau)$ is obtained by replacing each free occurrence of x in $\varphi(x)$ by τ.

DEFINITION: By a *fragment* of $L_{\omega_1\omega}$ we mean a set L_A of formulas such that:

(1) Every formula of L belongs to L_A;
(2) L_A is closed under $\neg$, $\exists x$, and finite disjunction.
(3) If $\varphi(x) \in L_A$ and τ is a term then $\varphi(\tau) \in L_A$.
(4) If $\varphi \in L_A$ then every subformula of φ belongs to L_A.

Note that for every set Ψ of formulas of $L_{\omega_1\omega}$, there is a least fragment $L_A \supset \Psi$. Moreover, if Ψ is countable then so is the least fragment containing Ψ.

From now on we shall assume that L_A is a countable fragment of $L_{\omega_1\omega}$. L_A is arbitrary but is held fixed to simplify notation.

Let C be a countable set of new constant symbols and form the first order language K by adding to L the constants $c \in C$. We shall let K_A be the set of all formulas obtained from formulas $\varphi \in L_A$ by replacing finitely many free variables by constants $c \in C$. Thus K_A is the least fragment of $K_{\omega_1\omega}$ which contains L_A. Note that each formula $\varphi \in K_A$ contains only finitely many $c \in C$.

To conserve notation we shall use the same symbol M for a model and for its set of elements. The models for K have the form $(M, a_c)_{c \in C}$ where M is a model for L and each constant $c \in C$ has the interpretation $a_c \in M$. The mapping $c \to a_c$ is called an *assignment* of C in M. We say that $(M, a_c)_{c \in C}$ is a *canonical* model for K iff the assignment $c \to a_c$ maps C onto M, i.e.,

$$M = \{a_c : c \in C\}.$$

Our starting point is the notion of a forcing property.

DEFINITION: A *forcing property* for the language L is a triple $\mathscr{P} = \langle P, \leqq, f \rangle$ such that

(i) $\langle P, \leqq \rangle$ is a partially ordered structure with a least element 0;
(ii) f is a function which associates with each $p \in P$ a set $f(p)$ of atomic sentences of K.
(iii) Whenever $p \leqq q$, $f(p) \subset f(q)$.
(iv) Let σ and τ be terms of K without variables and $p \in P$.

Then: If $(\tau = \sigma) \in f(p)$, then $(\sigma = \tau) \in f(q)$ for some $q \geqq p$. If $(\tau = \sigma)$, $\varphi(\tau) \in f(p)$, then $\varphi(\sigma) \in f(q)$ for some $q \geqq p$. For some $c \in C$ and $q \geqq p$, $(c = \tau) \in f(q)$.

The elements of P are called *conditions* of $\mathscr{P}$.

DEFINITION: The relation $p \Vdash \varphi$ in $\mathscr{P}$, read *p forces* φ, is defined recursively for conditions $p \in P$ and sentences $\varphi \in K_A$ as follows.

If φ is an atomic sentence, then $p \Vdash \varphi$ iff $\varphi \in f(p)$.
$p \Vdash \neg\varphi$ iff there is no $q \geqq p$ such that $q \Vdash \varphi$.
$p \Vdash \bigvee\Phi$ iff $p \Vdash \varphi$ for some $\varphi \in \Phi$.
$p \Vdash \exists x\varphi(x)$ iff $p \Vdash \varphi(c)$ for some $c \in C$.

We say that p *weakly forces* φ, in symbols $p \Vdash^w \varphi$, iff p forces $\neg\neg\varphi$.

The above definition is a generalization of the forcing studied in Robinson [30] for L and Barwise [3] for L_A. We describe their forcing as a first example.

Example 1.1. Let $\mathscr{M}$ be a class of models for L. The forcing property $\mathscr{P}(\mathscr{M})$ is described as follows. The conditions $p \in P$ are all finite sets of atomic and negated atomic sentences of K such that p

is satisfiable in some $M \in \mathscr{M}$. The relation $\leqq$ is the inclusion relation $\subset$ on P. For each $p \in P, f(p)$ is the set of all atomic sentences in p.

The next example is a generalization.

Example 1.2. Again let $\mathscr{M}$ be a class of models for L. Let Φ be a set of formulas of L_A which contains all atomic formulas and is closed under subformulas. We describe a forcing property $\mathscr{P}(\mathscr{M}, \Phi)$. $\Phi(C)$ is the set of all sentences $\varphi(c_1 \cdots c_n)$ of K_A where $\varphi(x_1 \cdots x_n) \in \Phi$. The conditions $p \in P$ are all finite subsets of $\Phi(C)$ which are satisfiable in some $M \in \mathscr{M}$. As before, $\leqq$ is $\subset$ and $f(p)$ is the set of all atomic sentences in p.

Some other types of forcing which can readily be expressed as forcing properties in our sense are set-theoretic forcing (see section 5) the forcing in number theory of Feferman [7], and the forcing in admissible sets of Barwise [1].

Here are some familiar facts about forcing.

LEMMA 1.3: *Let $\mathscr{P}$ be a forcing property and let $\varphi \in K_A$.*

(i) *p weakly forces φ iff for every $q \geqq p$ there is a condition $r \geqq q$ which forces φ.*
(ii) *If $p \leqq q$ and $p \Vdash \varphi$ then $q \Vdash \varphi$.*
(iii) *We cannot have both $p \Vdash \varphi$ and $p \Vdash \neg\varphi$.*
(iv) *If $p \Vdash \varphi$ then $p \Vdash^w \varphi$.*
(v) *$p \Vdash^w \neg\varphi$ iff $p \Vdash \neg\varphi$.*
(vi) *$p \Vdash \forall x\varphi(x)$ iff for all $c \in C$ and $q \geqq p$ there exists $r \geqq q$ such that $r \Vdash \varphi(c)$.*
(vii) *$p \Vdash \bigwedge\Phi$ iff for each $\varphi \in \Phi$ and $q \geqq p$ there exists $r \geqq q$ such that $r \Vdash \varphi$.*

The proof of (ii) is by an induction on the complexity of φ, while the other parts of the lemma are very easy.

DEFINITION: A subset $G \subset P$ is said to be *generic* iff

(i) $p \in G$ and $q \leqq p$ implies $q \in G$.
(ii) $p, q \in G$ implies that there exists $r \in G$ with $p \leqq r$ and $q \leqq r$.

(iii) For each sentence φ in K_A there exists $p \in G$ such that either $p \Vdash \varphi$ or $p \Vdash \neg\varphi$.

A generic set G is said to *generate* M iff M is a canonical model and every sentence φ of K_A which is forced by some $p \in G$ holds in M.

M is a *generic model* for a condition $p \in P$ iff M is generated by some generic set G which contains p.

By a *generic model* we mean a model M for the original language L such that for some assignment $\{a_c : c \in C\}$ in M, $(M, a_c)_{c \in C}$ is a generic model for the least condition in P. Thus every generic model is countable.

The reader should keep in mind that the notion of a generic model for p depends on the forcing property $\mathscr{P}$ and the fragment L_A. To be completely unambiguous we could use the following notation for a given forcing property $\mathscr{P}$ and fragment L_A:

$p \Vdash_{\mathscr{P}} \varphi$ for $p \Vdash \varphi$,
$p \Vdash^w_{\mathscr{P}} \varphi$ for $p \Vdash^w \varphi$,
G is $\mathscr{P}$, A-generic for G is generic,
G $\mathscr{P}$, A-generates M for G generates M,
M is $\mathscr{P}$, A-generic for M is generic.

In practice, however, $\mathscr{P}$ and L_A will be fixed throughout a given discussion and will not be carried in the notation.

GENERIC MODEL THEOREM: *If $\mathscr{P}$ is a forcing property and $p \in P$ then there is a generic model for p.*

This theorem follows from the two lemmas below.

LEMMA 1.4: *Every $p \in \mathscr{P}$ belongs to a generic set.*

Proof: Let $\varphi_0, \varphi_1, \varphi_2, \cdots$ be an enumeration of all sentences of K_A. Form a chain of conditions $p_0 \leqq p_1 \leqq \cdots$ in $\mathscr{P}$ as follows. Let $p_0 = p$. If $p_n \Vdash \neg\varphi_n$, let $p_{n+1} = p_n$. If p_n does not force $\neg\varphi_n$, choose $p_{n+1} \geqq p_n$ such that p_{n+1} forces φ_n. The set

$$G = \{q \in P : q \leqq p_n \text{ for some } n < \omega\},$$

is generic and contains p. ⊣

LEMMA 1.5: *Every generic set G of $\mathcal{P}$ generates a model. That is, there is a canonical model M such that every sentence forced by some $p \in G$ holds in M.*

Proof: Let T be the set of all sentences of K_A which are forced by some $p \in G$. Then T has the following properties for all sentences and all terms without variables in K_A.

(1) Exactly one of φ, $\neg\varphi$ belongs to T.
(2) $\bigvee\Phi \in T$ iff $\varphi \in T$ for some $\varphi \in \Phi$.
(3) $\exists x\varphi(x) \in T$ iff $\varphi(c) \in T$ for some $c \in C$.
(4) If $(\tau = \sigma) \in T$ then $(\sigma = \tau) \in T$.
If $(\tau = \sigma)$, $\varphi(\tau) \in T$ where $\varphi(x)$ is an atomic formula, then $\varphi(\sigma) \in T$.
For some $c \in C$, $(c = \tau) \in T$.

These are easily verified using the definition of a forcing property.

As in Henkin's proof of the completeness theorem, a model $(M, a_c)_{c \in C}$ of T can be constructed as follows. For c, $d \in C$, the equivalence relation $c \sim d$ is defined by

$$c \sim d \quad \text{iff} \quad (c = d) \in T.$$

Each $c \in C$ is assigned its equivalence class

$$a_c = \{d : c \sim d\},$$

and M is the set $M = \{a_c : c \in C\}$ of all equivalence classes. The function and relation symbols of L are interpreted in M in such a way that

$$F(a_{c_1}, \cdots, a_{c_n}) = a_c \quad \text{iff} \quad (F(c_1, \cdots, c_n) = c) \in T,$$
$$R(a_{c_1}, \cdots, a_{c_n}) \quad \text{iff} \quad (R(c_1, \cdots, c_n)) \in T.$$

The property (4) is used to show the unambiguity of this definition. Then properties (1)–(4) are used to prove by induction that for each sentence φ of K_A,

$$M \models \varphi \quad \text{iff} \quad \varphi \in T. \dashv$$

COROLLARY 1.6: *Let $\mathcal{P}$ be a forcing property and $p \in P$. Then $p \Vdash^w \varphi$ iff φ holds in every generic model for p. Thus if*

$$T = \{\varphi \in K_A : p \Vdash^w \varphi\} \quad \textit{and} \quad T \Vdash \psi \quad \textit{then} \quad p \Vdash^w \psi.$$

Proof: Let $p \Vdash^w \varphi$ and let M be a generic model for p. Then $p \Vdash \neg\neg\varphi$, hence $M \models \neg\neg\varphi$ and $M \models \varphi$.

Suppose p does not weakly force φ. Then for some $q \geqq p$, $q \Vdash \neg\varphi$. Let M be a generic model for q. Then $M \models \neg\varphi$ and M is a generic model for p. ⊣

Remember that to simplify notation we kept the countable fragment L_A fixed throughout our discussion. The proposition below is a summary of how the various notions are affected when the fragment L_A is changed. Each assertion follows easily from the definitions.

PROPOSITION 1.7: *Let L_A and L_B be countable fragments with $L_A \subset L_B$, and let $\mathscr{P}$ be a forcing property with respect to L.*

(i) *If $p \in P$ and $\varphi \in K_A$, then $p \Vdash \varphi$ with respect to L_A iff $p \Vdash \varphi$ with respect to L_B.*
(ii) *Every $\mathscr{P}$, B-generic set is a $\mathscr{P}$, A-generic set.*
(iii) *If $p \in P$, every $\mathscr{P}$, B-generic model for p is a $\mathscr{P}$, A-generic model for p.*
(iv) *Every $\mathscr{P}$, B-generic model is $\mathscr{P}$, A-generic.*

Our assumption that the fragment L_A and set C are countable was used in the proof of the Generic Model Theorem. The result can be generalized to apply to certain uncountable languages in the following way. Let κ be a regular cardinal and assume that L_A is a fragment of the language $L_{\kappa^+\omega}$ of power $\leqq\kappa$ and that the set C has power κ. Assume that the partially ordered structure $\langle P, \leqq\rangle$ has the property that for each $\alpha < \kappa$, any increasing sequence

$$p_0 \leqq p_1 \leqq \cdots \leqq p_\beta \leqq \cdots, \qquad \beta < \alpha,$$

in P has an upper bound. Then the Generic Model Theorem holds, that is, for every $p \in P$ there is a generic model. This uncountable form of the Generic Model Theorem is also essentially in [28].

The infinite forcing of Robinson [31] is an example of a forcing property in an uncountable language L_A.

2. THE OMITTING TYPES THEOREM

The Omitting Types Theorem was originally proved directly using the method of diagrams by Henkin [14] and Orey [27]. See also [11] and [19]. Here we shall give another proof which uses the Generic Model Theorem, and present some applications. We begin with a simple form of the result which deals with formulas having few quantifiers.

DEFINITION: By a *basic formula* we mean either an atomic formula or a negated atomic formula. Given a model M for L, a *finite piece* of M is a finite set p of basic sentences of K which is satisfiable by some assignment of C in M. If $\mathscr{M}$ is a class of models for L, then "satisfiable in $\mathscr{M}$" means "satisfiable in some $M \in \mathscr{M}$," and "finite piece of $\mathscr{M}$" means "finite piece of some $M \in \mathscr{M}$." The set P of all finite pieces of $\mathscr{M}$, with $\leq$ the inclusion relation $\subset$ and

$$f(p) = \{\varphi \in p : \varphi \text{ is atomic}\},$$

is the forcing property $\mathscr{P}(\mathscr{M})$ in Example 1.1 (studied by Robinson [30]). We shall call a generic model with respect to $\mathscr{P}(\mathscr{M})$ an $\mathscr{M}$*-generic model.*

A formula φ of L_A is said to be an $\forall\bigvee\exists$ *formula* iff φ is of the form

$$\forall x_1 \cdots \forall x_m \bigvee_{n<\omega} \exists y_1 \cdots \exists y_{i_n} (\varphi_{n1} \wedge \cdots \wedge \varphi_{nj_n})$$

where each φ_{ij} is a basic formula.

In what follows we shall sometimes write X^n for the set of all n-tuples of elements of a set X, and use the "vector notation" $\vec{x}$ for an n-tuple $\langle x_1, \cdots, x_n \rangle$.

THEOREM 2.1: *Let $\mathscr{M}$ be a class of models for L. An $\forall\bigvee\exists$ sentence*

$$\varphi = \forall x_1 \cdots \forall x_m \psi(x_1 \cdots x_m),$$

in L_A is true in all $\mathscr{M}$-generic models iff for every finite piece p of $\mathscr{M}$ and every m-tuple $\vec{c} \in C^m$, the set $p \cup \{\psi(\vec{c})\}$ is satisfiable in $\mathscr{M}$.

Proof: Let P be the set of all finite pieces of $\mathscr{M}$ and let

$$\psi(x_1 \cdots x_n) = \bigvee_{n<\omega} \exists y_1 \cdots \exists y_{i_n} (\varphi_{n1} \wedge \cdots \wedge \varphi_{nj_n}).$$

Each of the following statements is equivalent:

(1) φ holds in all $\mathscr{M}$-generic models.
(2) For all $\vec{c} \in C^m$, $\psi(\vec{c})$ holds in all $\mathscr{M}$-generic models for 0.
(3) For all $\vec{c} \in C^m$, $0 \Vdash^w \psi(\vec{c})$.
(4) For all $\vec{c} \in C^m$ and $p \in P$, there exists $q \supset p$ in P such that $q \Vdash \psi(\vec{c})$.
(5) For all $\vec{c} \in C^m$ and $p \in P$, there exists $q \supset p$ in P, $n < \omega$, and $\vec{d} \in C^{i_n}$ such that

$$q \Vdash \varphi_{n1}(\vec{c}, \vec{d}), \cdots, q \Vdash \varphi_{nj_n}(\vec{c}, \vec{d}).$$

(6) For all $\vec{c} \in C^m$ and $p \in P$ there exist $n < \omega$ and $\vec{d} \in C^{i_n}$ such that

$$p \cup \{\varphi_{nj}(c, \vec{d}): 1 \leqq j \leqq j_n\} \in P.$$

(7) For all $\vec{c} \in C^m$ and $p \in P$, $p \cup \{\psi(\vec{c})\}$ is satisfiable in $\mathscr{M}$.

The only nontrivial step is from (5) to (6). Let q be as in (5). q is a finite piece of some model $M \in \mathscr{M}$. If $\varphi_{n1}(\vec{c}, \vec{d})$ is atomic then it belongs to q. If $\varphi_{n1}(\vec{c}, \vec{d}) = \neg\psi$ where ψ is atomic, then $q \cup \{\psi\} \notin P$, so $q \cup \{\psi\}$ is not a finite piece of M. Therefore $q \cup \{\varphi_{n1}(,\vec{c}\, \vec{d})\}$ is a finite piece of M. Continuing, we see that

$$q \cup \{\varphi_{nj}(\vec{c}, \vec{d}): 1 \leqq j \leqq j_n\}$$

is a finite piece of M, and (6) follows. ⊣

The class of all models of a consistent set of $\forall\bigvee\exists$ sentences of L_A is called an $\forall\bigvee\exists$ *class* for L_A.

COROLLARY 2.2: *If $\mathscr{M}$ is an $\forall\bigvee\exists$ class for L_A, then every $\mathscr{M}$-generic model belongs to $\mathscr{M}$.*

The Omitting Types Theorem is a consequence of Theorem 2.1 which does not mention generic models.

BASIC OMITTING TYPES THEOREM: *Let $\mathscr{M}$ be an $\forall\bigvee\exists$ class and let*

$$\varphi_n = \forall x_1 \cdots \forall x_{m_n} \psi_n (\vec{x})$$

be a countable sequence of $\forall\bigvee\exists$ *sentences. Suppose that for each* n, *each finite piece* p *of* $\mathscr{M}$, *and each* m_n*-tuple* $\vec{c} \in C^{m_n}$, $p \cup \{\psi_n(\vec{c})\}$ *is satisfiable in* $\mathscr{M}$. *Then* $\mathscr{M}$ *contains a countable model in which each* φ_n *holds.*

Proof: We may assume the fragment L_A is large enough to contain each sentence φ_n. Then any $\mathscr{M}$-generic model belongs to $\mathscr{M}$ and satisfies each φ_n. ⊣

Here are two other corollaries which do not mention generic models.

COROLLARY 2.3: *If* $\mathscr{M}_i$, $i \in I$, *is a family of* $\forall\bigvee\exists$ *classes which all have exactly the same finite pieces, then* $\bigcap_{i \in I} \mathscr{M}_i \neq 0$.

Proof: For each $i \in I$, the $\mathscr{M}_i$-generic models are the same. ⊣

A model M is *locally finite* iff every finitely generated submodel of M is finite.

COROLLARY 2.4: *Suppose* L *has finitely many function and constant symbols. Let* $\mathscr{M}$ *be an* $\forall\bigvee\exists$ *class. Suppose that every finite piece of* $\mathscr{M}$ *is satisfiable in a finite submodel of some* $M \in \mathscr{M}$. *Then* $\mathscr{M}$ *has a locally finite model.*

Proof: For each n, let $\varphi(x_1 \cdots x_n)$ be the formula which says that each function or constant symbol applied to elements from $\{x_1, \cdots, x_n\}$ has a value equal to one of $x_1, \cdots, x_n$. $\varphi(x_1, \cdots, x_n)$ may be written in the form $\varphi_1 \vee \cdots \vee \varphi_{k_n}$ where each φ_i is a finite conjunction of atomic formulas. A model M is locally finite iff it satisfies the $\forall\bigvee\exists$ sentences

$$\theta_m = \forall x_1 \cdots \forall x_m \bigvee_{n<\omega} \bigvee_{i \leq k_n} \exists x_{m+1} \cdots \exists x_n \varphi_i(x_1, \cdots, x_n), m = 1,2,3,\cdots.$$

Let L_A be a countable fragment containing each θ_m. It follows from Theorem 2.1 that every $\mathscr{M}$-generic model is a model of each θ_m and hence is locally finite. ⊣

Theorem 2.1 and its corollaries can be generalized by replacing the set of basic formulas by another set of formulas. In what follows, Φ is assumed to be a set of formulas of L_A containing all atomic formulas and closed under subformulas. The classical case is where Φ is the set of all formulas of the finitary logic L.

Given a class $\mathscr{M}$ of models $\mathscr{P}(\mathscr{M}, \Phi)$ is the forcing property of

Example 1.2, where P is the set of all finite sets $p \subset \Phi(C)$ which are satisfiable in $\mathscr{M}$, $\leqq$ is the inclusion relation, and $f(p)$ is the set of all atomic sentences in p.

DEFINITION: We say that φ is an $\forall\vee\exists$ *formula over* Φ iff φ is a formula of L_A of the form

$$\forall x_1 \cdots \forall x_m \bigvee_{n<\omega} \exists y_1 \cdots \exists y_{i_n}(\varphi_{n1} \wedge \cdots \wedge \varphi_{nj_n})$$

where each φ_{ij} belongs to Φ.

LEMMA 2.5: *Let p be a condition in the forcing property $\mathscr{P}(\mathscr{M}, \Phi)$, and let $\varphi \in \Phi(C)$. If $p \models \varphi$ then $p \Vdash^w \varphi$, and if $p \Vdash^w \varphi$ then φ is consistent with p.*

Proof: The proof is by induction on the complexity of φ. The lemma is easy for atomic φ because in this case $p \Vdash \varphi$ iff $\varphi \in p$. Assume the result for all sentences ψ of lower complexity than φ. In each case below we assume that $p \leqq q$.

Case 1. $\varphi = \neg\psi$. Assume $p \models \neg\psi$. Then $q \models \neg\psi$, so ψ is not consistent with q, and hence q does not weakly force ψ. Then some $r \geqq q$ forces $\neg\psi$, whence $p \Vdash^w \neg\psi$. Now assume $p \Vdash^w \neg\psi$. Then $q \Vdash^w \neg\psi$, so q does not weakly force ψ. It follows that $q \not\models \psi$, so $\neg\psi$ is consistent with q and therefore with p.

Case 2. $\varphi = \bigvee\Psi$. Assume $p \models \bigvee\Psi$. Then $q \models \bigvee\Psi$ and q is satisfiable in $\mathscr{M}$, so for some $\psi \in \Psi$, $q \cup \{\psi\} = r$ is satisfiable in $\mathscr{M}$. Since $\psi \in \Phi(C)$, r is a condition and $r \geqq q$. We have $r \models \psi$, so $r \Vdash^w \psi$, and for some $s \geqq r$, $s \Vdash \psi$. Then $s \Vdash \bigvee\Psi$, hence $p \Vdash^w \bigvee\Psi$.

Assume $p \Vdash^w \bigvee\Psi$. Then some $r \geqq q$ forces $\bigvee\Psi$, so $r \Vdash \psi$ for some $\psi \in \Psi$. Thus ψ is consistent with r, and it follows that $\bigvee\Psi$ is consistent with p.

Case 3. $\varphi = \exists x\psi$. This is similar to Case 2. ⊣

Using the lemma, the proof of Theorem 2.1 and its corollaries may be generalized to give the following.

THEOREM 2.1 (General form): *Let $\mathscr{M}$ be a class of models for L and let*

$$\varphi = \forall x_1 \cdots x_m \psi(x_1 \cdots x_m)$$

be an $\forall\bigvee\exists$ *sentence over* Φ. *Then* φ *holds in all* $\mathscr{P}(\mathscr{M}, \Phi)$*-generic models iff for every m-tuple* $\vec{c} \in C^m$ *and every finite set* $p \subset \Phi(C)$ *which is satisfiable in* $\mathscr{M}$, *the set* $p \cup \{\psi(\vec{c})\}$ *is satisfiable in* $\mathscr{M}$.

The class of all models of a set of $\forall\bigvee\exists$ sentences over Φ is called an $\forall\bigvee\exists$ class over Φ.

COROLLARY 2.2 (General form): *If* $\mathscr{M}$ *is an* $\forall\bigvee\exists$ *class over* Φ *then every* $\mathscr{P}(\mathscr{M}, \Phi)$*-generic model belongs to* $\mathscr{M}$.

GENERAL OMITTING TYPES THEOREM: *Let* $\mathscr{M}$ *be an* $\forall\bigvee\exists$ *class over* Φ *and let*

$$\varphi_n = \forall x_1 \cdots \forall x_{m_n} \psi_n(\vec{x})$$

be a countable sequence of $\forall\bigvee\exists$ *sentences over* Φ. *Suppose that for each n, each finite set* $p \subset \Phi(C)$ *which is satisfiable in* $\mathscr{M}$ *and each* m_n*-tuple* $\vec{c} \in C^{m_n}$, *the set* $p \cup \{\psi_n(\vec{c})\}$ *is satisfiable in* $\mathscr{M}$. *Then* $\mathscr{M}$ *contains a countable model in which each* φ_n *holds.*

The book of Kreisel and Krivine contains extensions of the Omitting Types Theorem to uncountable languages (independently due to Chang) and to type theory. An extension to logic with extra quantifiers is obtained in [18].

We conclude this section with two applications of the Omitting Types Theorem to first order model theory.

DEFINITION: A mapping f from a model M into a model N for L is said to be an *elementary embedding* iff for every formula $\varphi(\vec{x})$ in L and every n-tuple $a_1, \cdots, a_n$ in M, we have

$$M \models \varphi(a_1 \cdots a_n) \quad \text{iff} \quad N \models \varphi(fa_1 \cdots fa_n).$$

M is an *elementary submodel* of N iff the identity map on M is an elementary embedding of M into N. Given a complete theory T in L, we say that M is a *prime model of* T iff M is elementarily embeddable in every model of T.

We shall characterize those theories T which have prime models. The characterization uses the notion of a complete formula.

A formula $\varphi(x_1 \cdots x_n)$ of L is said to be *complete* with respect to T iff φ is consistent with T and for every formula $\psi(x_1 \cdots x_n)$ of L, either $T \models \varphi \rightarrow \psi$ or $T \models \varphi \rightarrow \neg\psi$.

THEOREM 2.6: (Vaught [36].) *Let T be a complete theory in L. T has a prime model iff for every formula $\psi(x_1 \cdots x_n)$ of L which is consistent with T, there is a complete formula $\varphi(x_1 \cdots x_n)$ such that $T \models \varphi \to \psi$.*

Proof: Both directions use the Omitting Types Theorem where Φ is the set of all formulas of L.

Assume first that every consistent formula is implied by a complete formula with respect to T. For each n, let $\psi_n(x_1 \cdots x_n)$ be the disjunction of all complete formulas $\varphi(x_1 \cdots x_n)$ with respect to T, and let

$$\varphi_n = \forall x_1 \cdots \forall x_n \psi_n(x_1 \cdots x_n).$$

Then each φ_n is an $\forall\bigvee\exists$ sentence over Φ, and the class $\mathcal{M}$ of models of T is an $\forall\bigvee\exists$ class over Φ. Let $\vec{c} \in C^n$ and let $p(\vec{c}, \vec{d})$ be a finite subset of $\Phi(C)$ satisfiable in $\mathcal{M}$. Then $\exists\vec{y} \bigwedge p(\vec{x}, \vec{y})$ is consistent with T, so it is implied by a complete formula $\varphi(\vec{x})$. Therefore

$$\exists\vec{y} \bigwedge p(\vec{x}, \vec{y}) \wedge \psi_n(\vec{x})$$

is satisfiable in $\mathcal{M}$, whence $p(\vec{c}, \vec{d}) \cup \{\psi_n(\vec{c})\}$ is satisfiable in $\mathcal{M}$. By the Omitting Types Theorem, T has a countable model M in which each φ_n holds.

We claim that M is a prime model of T. To see this let N be any other model of T. Let $M = \{a_1, a_2, \cdots\}$ be an enumeration of M. For each n, the n-tuple $a_1, \cdots, a_n$ satisfies a complete formula $\theta_n(x_1 \cdots x_n)$ with respect to T. Moreover, we always have

$$T \models \theta_n \to \exists x_{n+1}\theta_{n+1},$$

because θ_n is complete and consistent with $\exists x_{n+1}\theta_{n+1}$. Let θ_0 be the true sentence. We may then choose elements $b_1, b_2, \cdots \in N$ such that for each n,

$$N \models \theta_n(b_1, \cdots, b_n).$$

Using completeness of the formulas θ_n again we see that the mapping $a_n \to b_n$ is a function and is an elementary embedding of M into N.

Now assume that T has a prime model M. Let $\psi(x_1 \cdots x_n)$ be consistent with T. Since T is complete, $T \models \exists x_1 \cdots \exists x_n \psi$, and therefore there is an n-tuple $\vec{a} \in M^n$ which satisfies ψ. Let $\Sigma(x_1 \cdots x_n)$

be the set of all formulas $\sigma(x_1 \cdots x_n)$ of L which are satisfied by $\vec{a}$ in M. Then for each formula $\varphi(x_1 \cdots x_n)$ of L, either $\varphi \in \Sigma$ or $(\neg\varphi) \in \Sigma$. Since M is prime, the set $\Sigma(x_1 \cdots x_n)$ is satisfied in every model of T. Therefore no model of T satisfies the $\forall\bigvee\exists$ sentence

$$\forall x_1 \cdots \forall x_n \bigvee_{\sigma \in \Sigma} \neg\sigma$$

over Φ. By the Omitting Types Theorem there is a finite set $p(\vec{c}, \vec{d}) \subset \Phi(C)$ such that $p(\vec{c}, \vec{d})$ is consistent with T but $p(\vec{c}, \vec{d}) \wedge \bigvee_{\sigma \in \Sigma} \neg\sigma(\vec{c})$ is not. Then for each $\sigma \in \Sigma$,

$$T \models \exists\vec{y} \bigwedge p(\vec{x}, \vec{y}) \to \sigma(\vec{x}).$$

It follows that $\exists\vec{y} \bigwedge p(\vec{x}, \vec{y})$ is a complete formula with respect to T. Since $\vec{a}$ satisfies $\psi(\vec{x})$ in M, we have $\psi(\vec{x}) \in \Sigma$, whence

$$T \models \exists\vec{y} \bigwedge p(\vec{x}, \vec{y}) \to \psi(\vec{x}).$$

Our proof is complete. ⊣

Vaught [36] has shown using a "back and forth" argument that if T has a prime model then it is unique up to isomorphism. The above proof shows that if T has a prime model, and if the fragment L_A is large enough to include the disjunction of all complete formulas $\psi(x_1 \cdots x_n)$ for each n, then each $(\mathcal{M}, \Phi)$-generic model is a prime model of T.

The next result is a Löwenheim-Skolem theorem for two cardinals. We let U be a unary predicate symbol of L. By a (κ, λ)-model we mean a model of power κ whose interpretation of U has power λ.

THEOREM 2.7: *Let T be a theory in L. If T has a (κ, λ)-model M where $\omega \leqq \lambda < \kappa$, then T has an (ω_1, ω)-model N. Moreover, there is a countable model M_0 and elementary embeddings $f: M_0 \to M$, $g: M_0 \to N$ such that g maps U^{M_0} onto U^N.*

The result was first proved without the "moreover clause" by Vaught in [24]. The stronger result above is in [17].

Sketch of Proof: The proof uses the basic results of Tarski and Vaught [35] on elementary submodels. By taking an elementary submodel of M, we may assume that M is a (λ^+, λ)-model. Add a new symbol $<$ to L and form a model $(M, <)$ where $<$ well orders M of order type λ^+. Let $(M_0, <_0)$ be a countable elementary submodel of M. We show that $(M_0, <_0)$ has a proper elementary extension $(M_1, <_1)$ in which all the new elements are greater than all $a \in M_0$ in the ordering $<_1$. This is done using the Omitting Types Theorem as follows.

Add to $L \cup \{<\}$ a new constant symbol c_a for each $a \in M_0$ and another constant symbol c. Let T' be the theory consisting of all sentences true in $(M_0, <_0, c_a)_{a \in M_0}$ together with the sentences

$$\{c_a < c : a \in M_0\}.$$

A sentence $\varphi(c)$ will be consistent with T' iff $\varphi(x)$ holds in $(M_0, <_0, c_a)_{a \in M_0}$ for arbitrarily large $x \in M_0$ under $<_0$. We can readily check that the hypotheses of the Omitting Types Theorem hold for the sentences

(1) $$\forall x \neg x < c_a \vee \bigvee_{b \in M_0} x = c_b, \qquad a \in M_0.$$

Therefore T' has a model $(M_1, <_1, c_a, c)_{a \in M_0}$ in which each sentence (1) holds. $(M_1, <_1)$ has the desired properties.

The same argument can be repeated ω_1 times forming a chain of proper elementary extensions $(M_\alpha, <_\alpha)$, $\alpha < \omega_1$. Let $(N, <)$ be the union of this chain. Then $(M_0, <_0)$ is an elementary submodel of $(N, <)$ so M_0 is an elementary submodel of N. N has power ω_1. Since U has power λ in M, U is "bounded" in $(M, <)$, i.e.,

$$(M, <) \models \exists x \forall y (U(y) \rightarrow y < x).$$

Therefore the above sentence holds in each $(M_\alpha, <_\alpha)$. Hence no new elements of U were added in forming the models $(M_\alpha, <_\alpha)$, and the interpretation of U is the same in N as in M_0. Since M_0 is countable, N is an (ω_1, ω)-model. ⊣

3. APPLICATIONS TO GROUP THEORY

The Basic Omitting Types Theorem can be applied to various branches of algebra. In this and the next section we discuss group theory and Archimedean fields.

The theory of groups is formulated in the language L with a binary operation $\cdot$, a unary operation $^{-1}$, and a constant symbol 1. The axioms are

$$\forall x \forall y \forall z (x \cdot y) \cdot z = x \cdot (y \cdot z)$$

$$\forall x (x \cdot 1 = x \wedge 1 \cdot x = x)$$

$$\forall x (x \cdot x^{-1} = 1 \wedge x^{-1} \cdot x = 1).$$

We may drop parentheses in products and give integer exponents x^n the usual meaning. Note that group theory is an $\forall\bigvee\exists$ theory in L, and in fact has just finitely many axioms all of which are universal. Thus if $\mathcal{M}$ is an $\forall\bigvee\exists$ class, so is the intersection of $\mathcal{M}$ and the class of all groups.

Several important notions in group theory can be expressed by $\forall\bigvee\exists$ sentences, and the Basic Omitting Types Theorem can be applied to them. We give three examples.

DEFINITION: A group M is said to be *periodic*, or *torsion*, iff every element of M has finite order, i.e.,

$$M \models \forall x \bigvee_{0<n<\omega} x^n = 1.$$

COROLLARY 3.1: *Let $\mathcal{M}$ be an $\forall\bigvee\exists$ class of groups. Suppose that for every finite piece p of $\mathcal{M}$ and every $c \in C$ there is a positive integer n such that $p \cup \{c^n = 1\}$ is satisfiable in $\mathcal{M}$. Then $\mathcal{M}$ contains a periodic group.*

Proof: Immediate from the Basic Omitting Types Theorem. ⊣

A group M is *divisible* iff each element has an nth root for all n, i.e.,

$$M \models \forall x \exists y (y^n = x), \qquad n = 1, 2, 3, \cdots.$$

COROLLARY 3.2: *Let $\mathcal{M}$ be an $\forall\bigvee\exists$ class of groups such that for each $n > 0$, $c \in C$, and finite piece p of $\mathcal{M}$, $p \cup \{\exists y y^n = c\}$ is satisfiable in $\mathcal{M}$. Then $\mathcal{M}$ contains a divisible group.*

DEFINITION: By a *word* $w(x_1 \cdots x_n)$ we mean a term of L in the variables $x_1, \cdots, x_n$. Given a set V of words and a group M, the *verbal subgroup* $V(M)$ generated by V is the subgroup of M generated by the set

(1) $\{w(c_1 \cdots c_n) : w(x_1 \cdots x_n) \in V \quad \text{and} \quad c_1, \cdots, c_n \in M\}.$

Two examples of verbal subgroups are the subgroup M^n generated by the word x^n, and the *commutator subgroup* $G^{(1)}$ generated by the word $x_1^{-1}x_2^{-1}x_1x_2$.

We shall let $\bar{V}$ denote the closure of V under products and inverses. Then for any group M,

$$V(M) = \{w(c_1 \cdots c_n) : w(x_1 \cdots x_n) \in \bar{V} \quad \text{and} \quad c_1, \cdots, c_n \in M\}.$$

COROLLARY 3.3: *Let $\mathcal{M}$ be an $\forall\bigvee\exists$ class of groups and let V be a set of words. Suppose that for each finite piece p of $\mathcal{M}$, there is a group $M_1 \in \mathcal{M}$ such that p is satisfiable in $V(M_1)$. Then $\mathcal{M}$ contains a group M such that $V(M) = M$.*

Proof: For each group M, we have $V(M) = M$ iff

$$M \models \forall x \bigvee_{w \in \bar{V}} \exists \vec{y} w(\vec{y}) = x.$$

The above is an $\forall\bigvee\exists$ sentence. Let p be a finite piece of $\mathcal{M}$ and let $c \in C$. For some $M_1 \in \mathcal{M}$, p is satisfiable in $V(M_1)$. Then

$$p \cup \left\{ \bigvee_{w \in \bar{V}} \exists \vec{y} w(\vec{y}) = c \right\}$$

is satisfiable in M_1. Therefore by the Basic Omitting Types Theorem, $\mathcal{M}$ contains a group M with $V(M) = M$. ⊣

The above three corollaries may also be viewed in the light of the Generic Model Theorem. Consider an $\forall\bigvee\exists$ class of groups $\mathcal{M}$. By Corollary 2.2, every $\mathcal{M}$-generic model belongs to $\mathcal{M}$ and is therefore a group, which we shall call an *$\mathcal{M}$-generic group*. Let us use Theorem 2.1 instead of the Omitting Types Theorem.

3.1*. If $\mathcal{M}$ is as in Corollary 3.1 and L_A contains the sentence

$$\forall x \bigvee_{0<n<\omega} x^n = 1,$$

then every $\mathcal{M}$-generic group is periodic.

3.2*. If $\mathscr{M}$ is as in Corollary 3.2, then every $\mathscr{M}$-generic group is divisible.

3.3*. If $\mathscr{M}$ is as in Corollary 3.3 and L_A contains the sentence

$$\forall x \bigvee_{w \in V} \exists \vec{y} w(\vec{y}) = x,$$

then every $\mathscr{M}$-generic group M_0 has the property $V(M_0) = M_0$.

One can continue along this line, showing that $\mathscr{M}$-generic groups have various properties given a sufficiently large fragment L_A.

In the literature there are several applications of forcing to algebraically closed groups. A group M is said to be *algebraically closed* iff for every finite set $s(x_1 \cdots x_m y_1 \cdots y_n)$ of basic formulas and all $c_1, \ldots, c_m \in M$, if

$$\exists y_1 \cdots \exists y_n \bigwedge s(c_1 \cdots c_m y_1 \cdots y_n)$$

holds in some extension group of M then it holds in M. These groups were defined and their existence proved by W. R. Scott [33]. They are in many ways analogous to algebraically closed fields.

A number of interesting properties of algebraically closed groups were obtained by Higman and the Neumanns, e.g., in [16], [26]. See also Macintyre [23]. For example, if M is an algebraically closed group, then:

(1) M is divisible.
(2) Every finite subset of M is contained in a two-generator subgroup of M.
(3) Any two elements $a, b \in M$ of the same order are conjugate in M, i.e., $b = x^{-1}ax$ for some $x \in M$.
(4) M is not finitely generated.
(5) M contains cyclic groups of all orders.

By a *generic group* we mean an $\mathscr{M}$-generic group where $\mathscr{M}$ is the class of all groups.

Let us generalize the notion of a generic group slightly. Consider a countable group M_0 and form the language $L(M_0)$ by adding a new constant symbol $\hat{a}$ for each $a \in M_0$. Let $\mathscr{M}_0$ be the class of all group extensions of M_0 with constants for each $a \in M_0$. Then $\mathscr{M}_0$ is an $\forall\bigvee\exists$ class of models for the language $L(M_0)$. By a *generic extension* of M_0 we mean an $\mathscr{M}_0$-generic model. Thus the generic groups are exactly the generic extensions of the one-element group.

THEOREM 3.4: (Robinson [30].) *Every generic extension of a countable group is algebraically closed.*

This gives another proof that every countable group can be extended to a countable algebraically closed group.

Proof: A generic extension M of M_0 is a member of $\mathscr{M}_0$ by Corollary 2.2, hence is an extension group of M_0. M is also countable. To prove that M is algebraically closed, let $c_1, \cdots, c_m \in C$ and let $p(\vec{x}, \vec{y})$ be a finite set of basic formulas such that $\exists\vec{y}\bigwedge p(\vec{c}, \vec{y})$ holds in some extension group of M. Let G be a generic set which generates $(M, a_c)_{c\in C}$. Let q be a condition in G, and let $d_1, \cdots, d_m$ be constants in C which do not occur in q or in $\{c_1, \cdots, c_m\}$. Since q holds in M, the set

$$p(\vec{c}, \vec{d}) \cup q$$

holds in some extension of M and therefore is a finite piece of an extension of M_0. It is easily seen using Lemma 1.3 that

$$p(\vec{c}, \vec{d}) \cup q \Vdash^{w} \exists\vec{y}\bigwedge p(\vec{c}, \vec{y}).$$

Therefore it is not the case that

$$q \nVdash \neg\exists\vec{y}\bigwedge p(\vec{c}, \vec{y}).$$

Since no $q \in G$ forces $\neg\exists\vec{y}\bigwedge p(\vec{c}, \vec{y})$, we have

$$(M, a_c)_{c\in C} \models \exists\vec{y}\bigwedge p(\vec{c}, \vec{y}),$$

so M is algebraically closed. ⊣

Let us now consider a finitely generated group M with generators $a_1, \cdots, a_n$. A *presentation* of M is a set P of equations in the constants $a_1, \cdots, a_n$ such that

(1) $M \models P$;

(2) For every equation φ in the constants $a_1, \cdots, a_n$, φ holds in M iff φ is provable from the set P and the group axioms.

It is a classical result of group theory that given a set P of equations in $a_1, \cdots, a_n$, there is a unique group M in the generators $a_1, \cdots, a_n$ which is presented by P. In particular, the empty set P presents the *free group* on n generators.

A group M generated by $a_1, \cdots, a_n$ is said to be *finitely presented* iff it is presented by a finite set of equations. M is said to be *recursively presented* iff M is presented by a recursive set of equations. M is said to be *recursively presented with a solvable word problem* iff the set of all equations in $a_1, \cdots, a_n$ true in M is recursive.

The following result uses forcing to obtain a characterization of recursively presented groups with a solvable word problem.

THEOREM 3.5: *A necessary and sufficient condition for a finitely generated group M to be recursively presented with a solvable word problem is that M is embeddable in every algebraically closed group.*

The necessity is due to Neumann [26] and Simmons [38], using [15]. We shall not give the proof here. The sufficiency is due to Macintyre [22] and is an application of forcing.

Proof of sufficiency: Let M be a finitely generated group with generators $a_1, \cdots, a_n$ and suppose M is not recursively presented with a solvable word problem. Let $\Phi(x_1 \cdots x_n)$ be the set of all basic formulas which are not satisfied by $a_1, \cdots, a_n$ in M. Then $\Phi(x_1 \cdots x_n)$ is not a recursive set. M is not embeddable in a group N iff

(1) $$N \models \forall x_1 \cdots \forall x_n \bigvee_{\varphi \in \Phi} \varphi(x_1 \cdots x_n).$$

Moreover,

(2) For each atomic formula $\varphi(x_1 \cdots x_n)$, exactly one of φ, $\neg\varphi$ belongs to Φ.

(3) Φ is not an r.e. set. (If Φ were r.e., its complement would also be r.e. by (2).)

Now let L_A be a countable fragment containing the formula $\bigvee\Phi$. We show that every generic group has the property (1). Let $c_1, \cdots, c_n \in C$ and let p be a finite piece of a group. Let $\Psi(x_1 \cdots x_n)$ be the set of all basic formulas $\psi(x_1 \cdots x_n)$ such that $p \cup \{\psi(c_1 \cdots c_n)\}$ is not a finite piece of a group. Then $\psi \in \Psi$ iff $\neg\psi(c_1 \cdots c_n)$ is provable from p and the group axioms. Therefore

(4) For each atomic formula ψ, at most one of ψ, $\neg\psi$ belongs to Ψ.

(5) Ψ is an r.e. set.

From (2)–(5) we see that Φ cannot be a subset of Ψ. Therefore there is a basic formula $\varphi \in \Phi - \Psi$, whence $p \cup \{\varphi(c_1 \cdots c_n)\}$ is a finite piece of a group. Then by Theorem 2.1, every generic group N satisfies (1), and by Theorem 3.4, N is algebraically closed. ⊣

4. SECOND ORDER NUMBER THEORY

In this section we shall work with a logic L which has a unary relation symbol N, a binary relation symbol $\in$, and constant symbols $0, 1, 2, \cdots$, plus a finite or countable list of other relation and function symbols. An *ω-model* M for L is a model in which N is the set of natural numbers, $M - N$ is a collection of subsets of N, and the symbols $\in$ and $0, 1, 2, \cdots$ are interpreted by themselves. (For convenience we assume that natural numbers are defined so that no element of M belongs to a natural number.) The class of models for L which are isomorphic to ω-models is an $\forall\bigvee\exists$ class given by the sentences

(1) $N(0), N(1), \cdots,$

(2) $0 \neq 1, 0 \neq 2, 1 \neq 2, \cdots,$

(3) $\forall x \forall y(x \in y \rightarrow N(x) \wedge \neg N(y)),$

(4) Axiom of extensionality:

$$\forall x \forall y(\forall z(z \in x \leftrightarrow z \in y) \rightarrow N(x) \vee N(y) \vee x = y),$$

(5) $\forall x \left(N(x) \rightarrow \bigvee_{n \in \omega} x = n \right).$

Note that each axiom (1)–(5) is an $\forall\bigvee\exists$ sentence, and all but (5) belong to L.

It is perhaps more natural to use a second order logic instead of the first order logic L, but then we would have to reformulate our general forcing theory in the framework of second order logic.

One of the original applications of the Omitting Types Theorem is the ω-completeness Theorem (Henkin [14] and Orey [27]).

ω-COMPLETENESS THEOREM: *Let T be a theory in L which contains the finitary axioms* (1)–(4) *above. Assume that for each formula $\varphi(x, \vec{y})$ of L, if*

$$T \models \varphi(n, \vec{y}) \quad \text{for} \quad n = 0, 1, 2, \cdots,$$

then

$$T \models \forall x(N(x) \to \varphi(x, \vec{y})).$$

Then T has an ω-model.

Proof: Let $p(x, \vec{y})$ be a finite set of formulas of L which is consistent with T. Let M be a model of T in which $c, \vec{d}$ satisfies $p(x, \vec{y})$. We wish to show that the set of formulas

$$(6) \qquad p(x, \vec{y}) \cup \{N(x) \to \bigvee_{n \in \omega} x = n\}$$

is consistent with T. If $\neg N(c)$, then $c, \vec{d}$ satisfies (6) in M. Assume $N(c)$. Then $T \not\models \forall x(N(x) \to \neg \bigwedge p(x, \vec{y}))$.

Therefore for some $n \in \omega$, $T \not\models \neg \bigwedge p(n, \vec{y})$. That is, $\bigwedge p(n, \vec{y})$ is consistent with T. Thus (6) is consistent with T. ⊣

In the following let Φ be a set of formulas of L_A which contains all basic formulas and is closed under subformulas.

We remark in passing that the ω-completeness Theorem also holds if T is a set of $\forall\bigvee\exists$ sentences over Φ and $\varphi(x, \vec{y})$ varies over formulas of the form $\neg\bigwedge p(x, \vec{y})$, where p is a finite subset of Φ. Our next application of the Omitting Types Theorem is similar to a result of Grilliot [10].

Definition: Let X be a set of natural numbers, and let M be a model for L. We say that X *belongs to* M iff there is an element $a \in M$ such that

$$X = \{n \in \omega : M \models n \in a\}.$$

Thus for ω-models M, X belongs to M iff $X \in M$.

Let T be a theory in L_A. We say that a set of formulas $p(x, \vec{y})$ *represents* X in T iff p is consistent with T and for all $n \in \omega$,

$$\text{if } n \in X \quad \text{then} \quad T \models \bigwedge p(x, \vec{y}) \to n \in x;$$

and

$$\text{if } n \notin X \quad \text{then} \quad T \models \bigwedge p(x, \vec{y}) \to n \notin x.$$

Theorem 4.1: *Let T be a consistent set of $\forall\bigvee\exists$ sentences over Φ, and let S be a countable collection of subsets of ω. Then there is a model M of T such that each $X \in S$ which belongs to M is represented in T by a finite set of formulas $p(x, \vec{y}) \subset \Phi$.*

Proof: We may assume that no $X \in S$ is represented in T by a finite $p \subset \Phi$. Then for each $X \in S$ and each finite set $p(x, \vec{y}) \subset \Phi$ consistent with T, the set of formulas

$$p(x, \vec{y}) \cup \left\{ \bigvee_{n \in X} \neg n \in x \vee \bigvee_{n \notin X} n \in x \right\}$$

is consistent with T. By the Omitting Types Theorem, T has a model M which satisfies for each $X \in S$ the sentence

$$\forall x \left(\bigvee_{n \in X} \neg n \in x \vee \bigvee_{n \notin X} n \in x \right).$$

Thus no $X \in S$ belongs to M. ⊣

Since the axioms for ω-models are $\forall\bigvee\exists$ sentences, we see that if T has an ω-model in Theorem 4.1 then M can be taken as an ω-model of T. Another immediate consequence of Theorem 4.1 is: If T is a consistent $\forall\bigvee\exists$ theory over Φ then every set $X \subset \omega$ which belongs to every model of T is representable in T by a finite set $p(x, \vec{y}) \subset \Phi$. It is of interest to consider special choices of Φ.

Example 4.2. If T is a consistent set of $\forall\bigvee\exists$ sentences, then every $X \subset \omega$ which belongs to every model of T is represented in T by a finite set of basic formulas.

Example 4.3. A Π_n formula in L is a formula of the form

$$\forall x_1 \exists x_2 \cdots Q x_n \varphi$$

where φ has only quantifiers restricted to N. Let T be a set of Π_{n+2} sentences of L with an ω-model. If X belongs to every ω-model of T then X is represented in T by a finite set of Π_n formulas.

Example 4.4. ([8], [11]). Let T be a consistent recursively axiomatizable theory in the finitary logic L. If a set X of natural numbers belongs to every model of T, then X is recursive.

Proof: X is representable in T by a finite set $p(x, \vec{y})$ of formulas of L. The set of consequences of $T \cup p(x, \vec{y})$ is recursively enumerable. Therefore both X and $\omega - X$ are recursively enumerable, so X is recursive. ⊣

Example 4.5. ([8], [11]). Let T be a Π_1^1 theory in L which has at least one ω-model. If a set X of natural numbers belongs to every ω-model of T, then X is hyperarithmetical.

Proof: Similar to Example 4.4. $\dashv$

COROLLARY 4.6: *Let T be a set of $\forall\bigvee\exists$ sentences over Φ such that every formula $\varphi \in \Phi$ which is satisfiable in some ω-model of T is satisfiable in every ω-model of T. Then for each ω-model M of T and each set $X \subset \omega$, the following are equivalent:*

(i) *X belongs to every ω-model of T.*
(ii) *There is a finite set $p(x, \vec{y}) \subset \Phi$ such that X is the unique element of M which satisfies $\exists\vec{y} \bigwedge p(x, \vec{y})$.*

Proof: We may assume T contains the axioms for ω-models.

Assume (i). Then X is represented in T by some finite $p(x, \vec{y}) \subset \Phi$. Since p is consistent with T it is satisfiable in M. Let X_0 satisfy $\exists\vec{y} \bigwedge p(x, \vec{y})$ in M. Then for each $n \in X$, $M \models n \in X_0$, and for each $n \notin X$, $M \models n \notin X_0$. Therefore $X = X_0$.

Assume (ii). Then $\exists x \exists\vec{y} \bigwedge p(x, \vec{y})$ holds in M and hence in every ω-model of T. If $n \in X$ then $\bigwedge p(x, \vec{y}) \wedge n \notin x$ is not satisfiable in M, so it is not satisfiable in any ω-model of T. Hence

$$T \models \bigwedge p(x, \vec{y}) \rightarrow n \in x.$$

Similarly for $n \notin X$. Therefore (i) holds. $\dashv$

The above results have analogues for ordered fields instead of second order number theory. This time we assume that L has the symbols $+$, $-$, $\cdot$, 0, 1, $<$, and finite or countably many additional function symbols. A model M for L is an ordered field iff $\langle M, +, -, \cdot, 0, 1\rangle$ is a field, $<$ is a linear ordering, and M satisfies

$$\forall x \forall y \forall z (x < y \rightarrow x + z < y + z),$$
$$\forall x \forall y \forall z (x < y \wedge 0 < z \rightarrow x \cdot z < y \cdot z).$$

The theory of ordered fields is a finite set of $\forall\bigvee\exists$ sentences of L. The integers $0, 1, 2, \cdots$ are defined as usual. An ordered field M is *Archimedean* iff it satisfies the $\forall\bigvee\exists$ sentence

$$\forall x \bigvee_{n<\omega} x < n.$$

Given an ordered field M, the *finite part of* M is the set of all $x \in M$ such that $-n < x < n$ for some positive integer n.

For simplicity we shall concentrate on basic formulas, which in L are just equalities and inequalities. We state two typical results.

COROLLARY 4.7: *Let* $\mathscr{M}$ *be an* $\forall\bigvee\exists$ *class of ordered fields. Suppose that every finite set of equalities and inequalities which has a solution in* $\mathscr{M}$ *has a solution in the finite part of some* $M \in \mathscr{M}$. *Then* $\mathscr{M}$ *contains an Archimedean ordered field.*

Proof: By the Basic Omitting Types Theorem. ⊣

Each Archimedean ordered field can be embedded in a unique way into the field of real numbers. We may therefore identify the elements of an Archimedean ordered field with real numbers.

COROLLARY 4.8: *Let* $\mathscr{M}$ *be an* $\forall\bigvee\exists$ *class of Archimedean ordered fields such that every finite set of equations and inequalities which has a solution in some member of* $\mathscr{M}$ *has a solution in every member of* $\mathscr{M}$. *Let* $M \in \mathscr{M}$ *and let* r *be a real number. Then the following are equivalent:*

(i) r *belongs to every member of* $\mathscr{M}$.
(ii) *For some finite set* $p(x, \vec{y})$ *of equations and inequalities,* r *is the unique real number in* M *such that* $p(r, \vec{y})$ *has a solution in* M.

Proof: Similar to Corollary 4.4. ⊣

For other applications to second order number theory see [7], [19], [21], and [37].

5. SET THEORY

We shall assume the reader is familiar with Zermelo-Fraenkel set theory, denoted by ZF. ZF with the axiom of choice is denoted by ZFC. Cohen's forcing is the chief tool for constructing models of ZF in the solution of consistency and independence problems. In this section we show how Cohen's forcing can be put into our present framework. We shall not start from the very beginning, but will simply state without proof the basic theorem which is needed to prove consistency results. After a couple of examples, we shall present for comparison some applications of the Omitting Types Theorem to models of set theory.

Rather than work with arbitrary models of ZF, we shall concentrate on standard models.

DEFINITION: A model $\langle M, E\rangle$ of ZF is said to be a *standard model* iff M is a transitive set, i.e., $x \in y \in M$ implies $x \in M$, and E is the $\in$ relation on M. We shall say that a set M is a standard model iff $\langle M, \in\rangle$ is a standard model.

Most relative consistency results for ZF have two versions, one for arbitrary models and one for standard models. Gödel [9] proved that:

(a) If ZF has a model, then ZFC has a model.
(b) If ZF has a standard model, then ZFC has a standard model.

There are now many other results which take the following two forms:

(a′) If ZF has a model, then ZF $\cup$ $\{\varphi\}$ has a model.
(b′) If ZF has a standard model, then ZF $\cup$ $\{\varphi\}$ has a standard model.

We shall discuss results of the form (b′). Let us first explain why standard models are easier to work with.

A Δ_0-*formula* is a formula built up from atomic formulas using $\neg$, $\vee$, and bounded quantifiers

$$(\exists x \in y)\varphi \;=\; \exists x(x \in y \;\wedge\; \varphi).$$

A formula $\varphi(\vec{x})$ is said to be *absolute* iff for every standard model M of ZF and all n-tuples $\vec{a}$ in M, we have

$$\langle M, \in\rangle \models \varphi(\vec{a}) \qquad \text{iff} \qquad \varphi(\vec{a}) \text{ is true.}$$

It can be shown by induction that every Δ_0 formula is absolute.

More generally, we say that $\varphi(\vec{x})$ is a Δ_1 *formula* in ZF iff there are Δ_0 formulas $\psi(\vec{x}, \vec{y})$ and $\theta(\vec{x}, \vec{y})$ such that

$$\text{ZF} \models \varphi(\vec{x}) \leftrightarrow \exists\vec{y}\psi(\vec{x}, \vec{y})$$

and

$$\text{ZF} \models \varphi(\vec{x}) \leftrightarrow \forall\vec{y}\theta(\vec{x}, \vec{y}).$$

It is easy to see that Δ_1 formulas of ZF are also absolute.

Most simple set-theoretic concepts are Δ_1 formulas in ZF, for example

$$x = \langle y, z \rangle, \qquad x \text{ is an ordinal,}$$

$$x = \cup y, \qquad x \text{ is a function.}$$

This makes standard models easy to deal with.

However, some statements, such as

$$x = \text{the power set of } y, \qquad x \text{ is a cardinal,}$$

cannot be expressed by Δ_1 formulas in ZF, or even ZFC.

DEFINITION: A relation $R(\vec{x})$ on M is said to be *definable* on M iff there is a formula $\theta(\vec{x})$ of L such that for all $\vec{a}$ in M, $R(\vec{a})$ iff $M \models \theta(\vec{a})$.

We need one more definition.

DEFINITION: Let M and N be standard models of ZF. M is said to be an *inner submodel* of N, and N an *outer extension* of M, iff $M \subset N$, and M and N have exactly the same ordinals.

The general plan of Cohen's method of proving results of the form (b′) is to let M be an arbitrary countable standard model of ZFC and show that M has an outer extension which is a model of $ZF \cup \{\varphi\}$. To get a countable standard model of ZFC in the first place, the following classical results are used:

LEMMA 5.1: (Gödel [9].)

(i) *Every standard model of* ZF *has a least inner submodel.*
(ii) *If M is a standard model of* ZF *with no proper inner submodel, then M satisfies the axiom of choice and the generalized continuum hypothesis.*

In fact, Gödel showed that M has no proper inner submodel iff M satisfies a certain sentence of ZF, the *axiom of constructibility*.

LEMMA 5.2: (Mostowski [25].) *A model $\langle M, E \rangle$ of* ZF *is isomorphic to a standard model iff the ordinals of $\langle M, E \rangle$ are well ordered by E.*

Corollary 5.3: *If* ZF *has a standard model then ZFC has a countable standard model.*

Proof: Let N be a standard model of ZF. Let M be the smallest inner submodel of N. Then M is a standard model of ZFC. By the Löwenheim-Skolem Theorem (Tarski-Vaught [35]), M has a countable elementary submodel M_0. Then M_0 is isomorphic to a standard model M_1, whence M_1 is a countable standard model of ZFC. $\dashv$

After these preliminaries we can state the fundamental result on set-theoretic forcing. Given a standard model M of ZFC, let $L(M)$ be the first order language with the symbol $\in$, a constant $\hat{a}$ for each $a \in M$, and an extra constant G. Form the expanded language $K(M)$ by adding a set $C = \{c_a : a \in M\}$ of new constants. All the symbols of $K(M)$ are elements of M.

Theorem 5.4: *Let M be a countable standard model of* ZFC, *and let $\langle P, \leqq \rangle \in M$ be a partially ordered structure with a least element. Then there is a forcing property $\mathscr{P} = \langle P, \leqq, f \rangle$ such that:*

(i) $p \Vdash (\hat{q} \in G)$ *iff* $q \leqq p$.
(ii) *For each formula $\varphi(\vec{x})$ of $L(M)$, the relation $p \Vdash \varphi(\vec{c})$, where $p \in P$ and $\vec{c}$ in C, is definable in $(M, a)_{a \in M}$.*
(iii) *Every generic model for $\mathscr{P}$ is isomorphic to an outer extension of M in which the axiom of choice holds and each $\hat{a}$ is interpreted by a.*

The above result is essentially due to Cohen [5]. For a more direct construction of the forcing property $\mathscr{P}$, see Shoenfield [34].

We remark that (iii) amounts to saying that the zero condition weakly forces all the axioms of ZFC and the infinite sentences

$$\forall x \left(x \in \hat{a} \rightarrow \bigvee_{b \in a} x = \hat{b} \right), \qquad a \in M,$$

$$\forall x \left(x \text{ is an ordinal} \rightarrow \bigvee_{a \in M} x = \hat{a} \right).$$

Here are two of Cohen's original applications of set-theoretic forcing:

THEOREM 5.5: *If* ZF *has a standard model, then* ZFC *has a standard model in which the axiom of constructibility fails.*

Proof: Let M be a countable standard model of ZFC. We show that M has a proper outer extension N which is a model of ZFC. Let P be the set of all finite partial functions p from ω into $\{0, 1\}$, and let $\leqq$ be the inclusion relation $\subset$ on P. Let $N = \langle N, \in, G_N, a\rangle_{a \in M}$ be a $\mathscr{P}$-generic model where $\mathscr{P}$ is the forcing property of Theorem 5.4 above. We need only show that $G_N \notin M$. Suppose $G_N = a \in M$. Then $p \Vdash \hat{a} = G$ for some $p \in P$. Choose n outside the domain of p and let $q = p \cup \{\langle n, 0\rangle\}$, $r = p \cup \{\langle n, 1\rangle\}$. Then $p \leqq q$, $p \leqq r$, but no $s \in P$ is $\geqq$ both q and r. Thus

$$q \Vdash \hat{q} \in G, \qquad r \Vdash \neg \hat{q} \in G.$$

Therefore if $q \in a$ then any generic model for r satisfies $\hat{a} \neq G$, and if $q \notin a$ then any generic model for q satisfies $\hat{a} \neq G$. It follows that p does not force $\hat{a} = G$ so G_N cannot belong to M. ⊣

THEOREM 5.6: *If* ZF *has a standard model, then* ZFC *has a standard model in which the continuum hypothesis* $2^{\omega} = \omega_1$ *fails.*

Proof: Let M be a countable standard model of ZFC. Let w_1 and w_2 be the cardinals ω_1 and ω_2 in the sense of M. (Thus w_1 and w_2 are countable ordinals which belong to M.) Let P be the set of all finite partial functions p on $w_2 \times \omega$ into $\{0, 1\}$ and let $\leqq$ be the inclusion relation on P. Let $N = \langle N, \in, G_N, a\rangle_{a \in M}$ be a $\mathscr{P}$-generic model. Then G_N induces a function F on w_2 into $S(\omega)$ defined by

$$F(\alpha) = \{n \in \omega : p(\alpha, n) = 1 \quad \text{for some} \quad p \in G_N\}.$$

Let F interpret the constant $\overline{F} \in C$ in N. For each $p \in P$ and $\alpha \neq \beta \in w_2$, we can find $n \in \omega$ and an extension $q \geqq p$ with $q(\alpha, n) = 0$, $q(\beta, n) = 1$. Therefore p cannot force $\overline{F}(\hat{\alpha}) = \overline{F}(\hat{\beta})$, and hence F is a 1-1 function.

To complete the proof we must show that ω, w_1, and w_2 have different cardinalities in N. This depends on the following combinatorial fact in M (we omit the proof):

(1) If X is a subset of P and no two elements of X have a common extension in P, then X is countable.

To show w_1 is not countable in N, suppose H is a function on ω into w_1 in N. We now work in M. Let p force $\overline{H}: \hat{\omega} \to \hat{w}_1$. For each $n \in \omega$, let $h(n)$ be the set of all $\alpha \in w_1$ such that some $q \geqq p$ forces $\overline{H}(\hat{n}) = \hat{\alpha}$. For each $\alpha \in h(n)$, choose a $q_\alpha \geqq p$ forcing $\overline{H}(\hat{n}) = \hat{\alpha}$. Then for $\alpha \neq \beta \in h(n)$, q_α and q_β have no common extension in P. Therefore by (1) there are only countably many q_α's, so $h(n)$ is countable. Then $\bigcup_{n \in \omega} h(n)$ is countable. Therefore there exists $\alpha \in w_1 - \bigcup_{n \in \omega} h(n)$. Then no $q \geqq p$ forces $(\exists n)\overline{H}(n) = \hat{\alpha}$. Therefore $p \Vdash \forall n \overline{H}(n) \neq \hat{\alpha}$. Returning to N, H maps ω properly into w_1, so w_1 is uncountable in N.

A similar proof shows that w_2 has cardinality greater than w_1 in N. The proof is complete. $\dashv$

The construction of a standard model of ZF in which the axiom of choice fails can also be done along these lines. The desired model turns out to be an inner submodel of a generic model N found using Theorem 5.4 (see [34]). Forcing in set theory has been developed far beyond Cohen's original results; a useful reference is Jech's book.

We now present some applications of the Omitting Types Theorem to models of set theory. These applications appear to be useless for consistency results because they give elementary extensions of a given model of ZF. However, they give information about what kinds of elementary extensions are possible.

We first note that no standard model of ZF has a proper outer elementary extension. For if N is an outer elementary extension of M, then for each $b \in N$ there is an ordinal $\alpha \in N$ and an element $a \in N$ such that $N \models b \in R(\alpha) \wedge a = R(\alpha)$; but then $\alpha \in M$, $a \in M$, $b \in a$, whence $b \in M$. Let us weaken the notion of an outer extension slightly.

Definition: Let M, N be standard models of ZF. An embedding $f: M \to N$ is an *outer embedding* iff M and N have the same ordinals and $f(\alpha) = \alpha$ for each ordinal $\alpha \in M$.

Thus N is an outer extension of M iff the identity map is an outer embedding. We shall characterize those M which have proper outer elementary embeddings.

Definition: In ZF, the classes W_α, α an ordinal, are defined as follows:

$x \in W_0$ iff x is well orderable.
$x \in W_{\alpha+1}$ iff $x = \bigcup y$ for some y such that $y \in W_0$ and $y \subset W_\alpha$.
If α is a limit ordinal, $x \in W_\alpha$ iff $x \in W_\beta$ for some $\beta < \alpha$.
Define $W = \bigcup_\alpha W_\alpha$.

The axiom of choice is equivalent to the statement $W_0 = V$ (the class of all sets). In general we have $W_\beta \subset W_\alpha$ when $\beta < \alpha$. It is known that if ZF has a standard model, then so does ZF $\cup \{W \neq V\}$ (see [12]). The theorem below is proved in [19].

THEOREM 5.7: *Let M be a countable standard model of ZF. The following are equivalent:*

(i) $M \models W \neq V$.
(ii) *M has a proper outer elementary embedding into some standard model N.*

Proof: First assume $M \models V = W$ and let $f\colon M \to N$ be an outer elementary embedding. We shall show by a double induction that f is the identity mapping, whence f is not proper. Let α be an ordinal of M and assume that $f(x) = x$ whenever $M \models x \in R(\alpha)$. Then let β be an ordinal of M and assume that for all $\gamma < \beta$, $f(y) = y$ whenever $M \models y \in R(\alpha + 1) \wedge y \in W_\gamma$. Under these assumptions we can prove that $f(z) = z$ whenever $M \models z \in R(\alpha + 1) \wedge z \in W_\beta$. This shows that $f(x) = x$ for all $x \in W$. Hence (ii) implies (i).

The proof that (i) implies (ii) uses the Omitting Types Theorem with Φ the set of all formulas of $L(M)$. Choose an element $b \in M$ with $M \models b \notin W$. Let T be a theory in $L(M)$ which contains every sentence true in $\langle M, \in, a\rangle_{a \in M}$ plus all sentences $\varphi(G)$ such that

$$M \models \{y \in b\colon \neg\varphi(y)\} \in W.$$

Then $T \models G \in \hat{b}$, and $\theta(G)$ is consistent with T iff

$$M \models \{y \in b\colon \theta(y)\} \notin W.$$

Let *ord* be the set of all ordinals of M and consider the sentence

$$(1) \qquad \forall x \left(ord\, x \to \bigvee_{\alpha \in ord} x = \hat{\alpha} \right).$$

Let $\theta(G, \vec{c}, d)$ be consistent with T, where $\vec{c}$, d are in C, and suppose

$$\theta(G, \vec{c}, d) \wedge \neg ord(d)$$

is not consistent with T. Then $\theta(G, \vec{c}, d) \wedge ord(d)$ is consistent with T, whence

$$M \models \{y \in b: \exists u \exists \vec{z} \theta(y, \vec{z}, u) \wedge ord(u)\} \notin W.$$

But then there exists $\alpha \in ord$ such that $M \models \{y \in b: \exists \vec{z} \theta(y, \vec{z}, \alpha)\} \notin W$, so $\theta(G, \vec{c}, d) \wedge d = \hat{\alpha}$ is consistent with T. It now follows from the Omitting Types Theorem that T has a model $\langle N_0, \in, a, G\rangle_{a \in M}$ in which (1) holds. Then N_0 is an elementary extension with exactly the same ordinals as M. Since $T \models G \in \hat{b}$, $N_0 \models G \in b$. Moreover, for each $a \in b$, $T \models G \neq \hat{a}$ so $N_0 \models G \neq a$. Therefore N_0 is a proper extension of M. Finally since the ordinals of N_0 are well ordered, N_0 is isomorphic to a standard model N, and this isomorphism induces a proper outer elementary embedding of M into N. ⊣

We remark that in the above proof, $(N_0, a, G)_{a \in M}$ is a generic model for the forcing property of all finite sets of sentences of $K(M)$ consistent with T. The proof shows that if $M \models S(\omega) \notin W$, then there is an outer embedding $f: M \to N$ in which N contains a new subset of ω.

Another use of the Omitting Types Theorem is to construct end elementary extensions of models of ZF.

Definition: Let $\langle M, E\rangle$ be a countable model of ZF. An extension $\langle N, F\rangle$ of $\langle M, E\rangle$ is said to be an *end extension* iff for all $b \in N$ and $a \in M$, if bFa then $b \in M$.

Theorem 5.8: (Keisler-Morley [20].) *Every countable model of* ZF *has a proper end elementary extension.*

Proof: We argue as in the preceding theorem, except that the theory T in $L(M)$ contains all sentences true in $\langle M, E, a\rangle_{a \in M}$ plus all sentences $\varphi(G)$ where $\langle M, E, a\rangle_{a \in M}$ satisfies

$$(\exists x)(\forall y)(ord(y) \wedge y \notin x \to \varphi(y)).$$

Using the Omitting Types Theorem we obtain a model

$$\langle N, F, a, G\rangle_{a \in M}$$

of T in which each of the infinite sentences

$$\forall x \left(x \in \hat{a} \to \bigvee_{b \in M} x = \hat{b} \right), \qquad a \in M$$

holds. Then $\langle N, F\rangle$ is an end elementary extension of $\langle M, E\rangle$. Since no element of M satisfies all of the sentences $\varphi(G)$, G must belong to $N - M$, whence $\langle N, F\rangle$ is a proper extension of $\langle M, E\rangle$. ⊣

The above theorem can be repeated ω_1 times to get an end elementary extension $\langle N, F\rangle$ of power ω_1.

The theorem does not give a standard model N, that is, it is not true that every countable standard model M of ZF has a standard proper end elementary extension N.

On the other hand, by modifying the proof it can be shown that every countable model $\langle M, E\rangle$ of ZF has an end elementary extension $\langle N, F\rangle$ which has an infinite decreasing sequence of ordinals $\cdots Fb_3Fb_2Fb_1$. Taking M standard we can get an ω-model N which is not isomorphic to any standard model.

We conclude with an end extension theorem of Barwise [2] whose method of proof lies somewhere between the Omitting Types Theorem and Cohen's forcing. Let ZFL be ZF plus the axiom of constructibility.

THEOREM 5.9: *Every countable standard model M of* ZF *has an end extension which is a model of* ZFL.

We shall indicate the main steps of the proof. This time let $L(M)$ be the language L plus a constant $\hat{a}$ for each $a \in M$, and form $K(M)$ by adding new constants c_a, $a \in M$. Let $L_A(M)$ be the fragment generated by the infinitary sentences

$$\delta_d = \bigwedge_{a \in d} \forall x \left(x \in \hat{a} \rightarrow \bigvee_{b \in a} x = \hat{b} \right), \qquad d \in M.$$

We need three lemmas:

LEMMA A: (i) *The set of all sentences of $K_A(M)$ is definable in M.* (ii) *The set of all sentences φ of $K_A(M)$ such that φ is consistent with* ZFL *is definable in M by a formula $\theta(x)$.*

The proof of this lemma is outside the scope of this paper. It uses the completeness theorem of Barwise [1] for infinitary logic (see [19]).

Consider the set P of all finite sets p of sentences in $K_A(M)$ such that for each $a \in M$, $p \cup \{\delta_a\}$ is consistent with ZFL.

LEMMA B: *$\langle P, \leqq, f\rangle$ is a forcing property where $\leqq$ is $\subset$ and $f(p)$ is the set of all atomic sentences in p.*

The only difficulty is to show that P is non-empty (whence the empty set is the least element of P). This uses a sharpened form of Lemma A, namely that the formula $\theta(x)$ of Lemma A is Π_1 and works for every countable standard M. It also uses the Shoenfield absoluteness lemma.

LEMMA C: *Every generic model with respect to $\mathscr{P}$ is an end extension of M and a model of* ZFL.

Proof: We first show that for each sentence φ of $K_A(M)$ and $p \in P$:

(1) If $p \models \varphi$ then $p \Vdash^w \varphi$.
(2) If $p \Vdash^w \varphi$ then $p \cup \{\varphi\} \in P$.

This is done by an induction on the complexity of φ which is like the proof of Lemma 2.5. The only case where new difficulties arise is Part (1) when $\varphi = \bigvee\Psi$.

Assume $p \models \bigvee\Psi$. Let $q \geqq p$. Since $L_A(M)$ is the least fragment containing all the sentences δ_a, the set Ψ belongs to M. We shall show that for some $\psi \in \Psi$, $q \cup \{\psi\} \in P$. Assume not. Then for each $\psi \in \Psi$ there exists $a \in M$ such that $q \cup \{\psi, \delta_a\}$ is not consistent with ZFL. Using the formula $\theta(x)$ from Lemma A,

$$M \models (\forall\psi \in \Psi)(\exists a)\, \neg\theta(q \cup \{\psi, \delta_a\}).$$

Note that if $a \supset b$ then δ_a implies δ_b. Thus using the axiom of replacement in M,

$$M \models (\exists a)(\forall\psi \in \Psi)\, \neg\theta(q \cup \{\psi, \delta_a\}).$$

Hence for all $\psi \in \Psi$, $q \cup \{\psi, \delta_a\}$ is inconsistent with ZFL. Then $q \cup \{\bigvee\Psi, \delta_a\}$ is inconsistent with ZFL, contradicting $q \cup \{\bigvee\Psi\} \in P$. So we have $r = q \cup \{\psi\} \in P$ for some $\psi \in \Psi$. By induction, $r \Vdash^w \psi$. Therefore $p \Vdash^w \bigvee\Psi$. This proves (1).

To finish the proof we note that each axiom of ZFL and each sentence δ_a, $a \in M$, is weakly forced by the empty condition. This is because if $p \in P$, $p \cup \{\varphi\} \in P$ for each of these sentences φ. ⊣

For other examples of the construction of models of set theory using the Omitting Types Theorem see [19].

All the results in this section have also been stated and proved for arbitrary countable models in place of countable standard models.

BACKGROUND READING

Chang, C. C., and H. J. Keisler. *Model Theory.* North Holland, 1973.

Cohen, P., *Set Theory and the Continuum Hypothesis.* Benjamin, 1966.

Jech, T., *Lectures in Set Theory.* Springer-Verlag, 1971.

Keisler, H. J., *Model Theory for Infinitary Logic.* North Holland, 1971.

Kreisel, G., and J. Krivine, *Elements of Mathematical Logic.* North Holland, 1967.

Neumann, H., *Varieties of Groups.* Springer-Verlag, 1967.

Rosser, J. B., *Simplified Independence Proofs.* Academic Press, 1969.

Shoenfield, J., *Mathematical Logic.* Addison-Wesley, 1967.

REFERENCES

1. Barwise, J., *Infinitary logic and admissible sets.* Doctoral thesis, Stanford University, 1967.
2. ———, *Infinitary Methods in the Model Theory of Set Theory.* Logic Colloquium '69, R. Gandy and C. M. E. Yates, eds. North Holland, 1971, 53–66.
3. ———, "Notes on forcing and countable fragments," mimeographed 1970 (unpublished).
4. Barwise, J., and A. Robinson, "Completing theories by forcing," *Ann. Math. Logic*, **2** (1970), 119–142.
5. Cohen, P., "The independence of the continuum hypothesis," *Proc. Nat. Acad. Sci.*, **50** (1963), 1143–1148 and **51** (1964), 105–110.
6. ———, *Set Theory and the Continuum Hypothesis.* Benjamin, 1966.
7. Feferman, S., "Some applications of the notions of forcing and generic sets," *Fund. Math.*, **56** (1965), 325–345.
8. Gandy, R. O., G. Kreisel, and W. W. Tait, "Set existence," *Bull. de l'Acad. Polon. des Sci.*, **8** (1960), 577–582 and **9** (1961), 881–882.
9. Gödel, K., *The Consistency of the Axiom of choice and the Generalized Continuum Hypothesis.* Princeton, 1940.

10. Grilliot, T., "Omitting types: application to recursion theory," *J. Symb. Logic*, **37** (1972), 81–89.

11. Grzegorczyk, A., A. Mostowski, and C. Ryll-Nardzewski, "Definability of sets in models of axiomatic theories," *Bull. Acad. Polon. Sci.*, **9** (1961), 163–167.

12. Halpern, J. D., and A. Lévy, " The boolean prime ideal theorem does not imply the axiom of choice," *Axiomatic Set Theory*, part 1. Providence, R.I., 1971, 83–134.

13. Henkin, L., "The completeness of the first-order functional calculus," *J. Symb. Logic*, **14** (1949), 159–166.

14. ———, "A generalization of the concept of ω-consistency," *J. Symb. Logic*, **19** (1954), 183–196.

15. Higman, G., "Subgroups of finitely presented groups," *Proc. Roy. Soc. London*, **262** (1961), 455–475.

16. Higman, G., B. H. Neumann, and H. Neumann, "Embedding theorems for groups," *J. London Math. Soc.*, **24** (1949), 247–254.

17. Keisler, H. J., "Some model-theoretic results for ω-logic," *Israel J. Math.*, **4** (1966), 249–261.

18. ———, "Logic with the quantifier 'there exists uncountably many'," *Ann. Math. Logic*, **1** (1970), 1–93.

19. ———, *Model Theory for Infinitary Logic*. North Holland, 1971.

20. Keisler, H. J., and M. Morley, "Elementary extensions of models of set theory," *Israel J. Math.*, **6** (1968), 49–65.

21. Kreisel, G., *Model-Theoretic Invariants. The Theory of Models*, J. Addison, L. Henkin, and A. Tarski, eds. North Holland, 1965, 190–205.

22. Macintyre, A., "Omitting quantifier-free types in generic structures," *J. Symb. Logic*, **37** (1972), 512–520.

23. ———, "On algebraically closed groups," *Ann. of Math.*, **96** (1972), 53–97.

24. Morley, M., and R. Vaught, "Homogeneous universal models," *Math. Scand.*, **11** (1962), 37–57.

25. Mostowski, A., "An undecidable arithmetical statement," *Fund. Math.*, **36** (1949), 143–164.

26. Neumann, B. H., "The isomorphism problem for algebraically closed groups," *Word Problems*, North Holland, 1973, 553–562.

27. Orey, S., "On ω-consistency and related properties," *J. Symb. Logic*, **21** (1956), 246–252.

28. Rasiowa, H., and R. Sikorski, *The mathematics of metamathematics*, Warsaw 1963.

29. Robinson, A., *On the Metamathematics of Algebra*. North Holland, 1951.

30. ———, "Forcing in model theory," *1st. Nat. Alta Math., Symposia Math.*, **5** (1970), 69–82.

31. ———, "Infinite forcing in model theory," *Proc. of the Second Scandinavian Logic Symposium*, North Holland, 1971, 317–340.

32. Scott, D., and R. Solovay. "Boolean-valued models for set theory," *Axiomatic Set Theory*, part 2, to appear.

33. Scott, W. R., "Algebraically closed groups," *Proc. Amer. Math. Soc.*, **2** (1951), 118–121.

34. Shoenfield, J., "Unramified forcing," *Axiomatic Set Theory*, part 1. Providence, R.I., 1971, 357–382.

35. Tarski, A., and R. Vaught, "Arithmetical extensions of relational systems," *Comp. Math.*, **13** (1957), 81–102.

36. Vaught, R., "Denumerable models of complete theories," *Infinitistic Methods*, New York and Warsaw, 1961, 303–321.

37. Mostowski, A., and Y. Suzuki, "On ω-models which are not β-models," *Fund. Math.*, **65** (1968), 83–93.

38. Simmons, H., "The word problem for absolute presentations," *J. London Math. Soc.*, to appear.

University of Wisconsin, Madison

MODEL THEORY AS A FRAMEWORK FOR ALGEBRA

Abraham Robinson[1]

1. INTRODUCTION

Model Theory is concerned with the interconnections between sentences, or sets of sentences formulated in a specified language, on one hand, and the mathematical structures in which these sentences are interpreted, on the other hand. It thus relates naturally to Algebra which is, or at least has been, for the past half century, the study of a number of particular axiomatic systems, e.g., for the notions of a group, of a field, of a Lie algebra, and of the corresponding classes of mathematical structures. By contrast, other branches of Mathematics, such as Number Theory, Set Theory, Classical Analysis, and even Functional Analysis are related to particular mathematical structures, i.e., *the* system of natural numbers, or *the* universe of sets, or *the* fields of real or complex

---

1. The research incorporated in this paper was supported in part by the National Science Foundation Grant No. GP18728.

then

$$T \models \forall x(N(x) \rightarrow \varphi(x, \vec{y})).$$

Then T has an ω-model.

Proof: Let $p(x, \vec{y})$ be a finite set of formulas of L which is consistent with T. Let M be a model of T in which $c, \vec{d}$ satisfies $p(x, \vec{y})$. We wish to show that the set of formulas

$$(6) \qquad p(x, \vec{y}) \cup \{N(x) \rightarrow \bigvee_{n \in \omega} x = n\}$$

is consistent with T. If $\neg N(c)$, then $c, \vec{d}$ satisfies (6) in M. Assume $N(c)$. Then $T \not\models \forall x(N(x) \rightarrow \neg \bigwedge p(x, \vec{y}))$.

Therefore for some $n \in \omega$, $T \not\models \neg \bigwedge p(n, \vec{y})$. That is, $\bigwedge p(n, \vec{y})$ is consistent with T. Thus (6) is consistent with T. ⊣

In the following let Φ be a set of formulas of L_A which contains all basic formulas and is closed under subformulas.

We remark in passing that the ω-completeness Theorem also holds if T is a set of $\forall\bigvee\exists$ sentences over Φ and $\varphi(x, \vec{y})$ varies over formulas of the form $\neg\bigwedge p(x, \vec{y})$, where p is a finite subset of Φ. Our next application of the Omitting Types Theorem is similar to a result of Grilliot [10].

DEFINITION: Let X be a set of natural numbers, and let M be a model for L. We say that X *belongs to* M iff there is an element $a \in M$ such that

$$X = \{n \in \omega : M \models n \in a\}.$$

Thus for ω-models M, X belongs to M iff $X \in M$.

Let T be a theory in L_A. We say that a set of formulas $p(x, \vec{y})$ *represents* X in T iff p is consistent with T and for all $n \in \omega$,

$$\text{if } n \in X \quad \text{then} \quad T \models \bigwedge p(x, \vec{y}) \rightarrow n \in x;$$

and

$$\text{if } n \notin X \quad \text{then} \quad T \models \bigwedge p(x, \vec{y}) \rightarrow n \notin x.$$

THEOREM 4.1: *Let T be a consistent set of* $\forall\bigvee\exists$ *sentences over* Φ, *and let S be a countable collection of subsets of* ω. *Then there is a model M of T such that each* $X \in S$ *which belongs to M is represented in T by a finite set of formulas* $p(x, \vec{y}) \subset \Phi$.

numbers. Whether or not one believes that the definite article which is italicized here has any absolute meaning in this connection, or whether one accepts it only as a convention, depends on one's philosophical convictions. In any case, the uniqueness (categoricity) of these systems is at the base of the several theories just mentioned. As it happens, the model theoretic approach has affected these domains also, by pointing up the existence of mathematical structures which, in each case, share any specified aggregate of formal properties with the intended or *standard* model of the theory under consideration. However, this aspect of Model Theory will be touched upon only in passing. Here, it will be our main purpose to show how certain basic facts and notions of Algebra, for example the notion of an algebraically closed field, can be placed and generalized within the framework of Model Theory. Many of the results described here are classical, i.e., more than ten years old (compare [7] which contains further references), but some significant developments of recent date have also been taken into account. Preceding the main discussion, several simple applications of Model Theory are presented as appetizers.

2. SOME BASIC NOTIONS AND RESULTS

Throughout this paper, we shall confine ourselves to the formal language of the first order, or lower, predicate calculus. Except for explicit mention, our arguments and results are independent of our particular choice among the variants of this calculus. Thus, we may or may not include function symbols among the atomic symbols of our language; we may regard the relation symbol of equality as denoting a particular equivalence relation with the substitutivity property, or we may take it to indicate identity (i.e., $a = b$ means that a and b denote the same object in the structure under consideration); and we may or may not include individual constants in the vocabulary of our axioms.

For example, the theory of fields can be formulated in terms of the identity $=$, the two-place function symbols $+$ and $\cdot$ and the individual constant 0 (with the usual interpretation); or in terms of the relation symbols $E(x, y)$ (for $x = y$) and $S(x, y, z)$, $P(x, y, z)$ (for

$x + y = z$ and $xy = z$) and without any individual constants (or again, with 0). Similar remarks apply to other theories. However, our choice of *vocabulary* may affect the particular syntactical form in which the axioms of the theory can be formulated. Whenever we talk of a *theory* T we shall mean a deductively closed set of sentences. By contrast a *set of axioms* K *for* T is a subset of T such that T is the deductive closure of K.

The class of *universal formulae* of a language (including in particular *universal sentences*) is defined inductively, as follows. (i) Any quantifier free formula is universal; (ii) the conjunction and the disjunction of two universal formulae are universal; (iii) a formula obtained from a universal formula by universal quantification is universal; (iv) no other formula is universal, i.e., the class of universal formulae is defined as the smallest class satisfying (i), (ii), and (iii). Similarly, the class of *existential* formulae is defined by (i)–(iv), the word *existential* replacing the word *universal* everywhere. Every universal (existential) formula is logically equivalent to a universal (existential) formula in prenex normal form expressed in the same vocabulary. An existential formula is called *primitive* if it is in prenex normal form and if its *matrix* (its quantifier free part) consists of a conjunction of *basic* formulae. A *basic formula* is an *atomic* formula or a negation of an atomic formula. A theory T is called *universal* (*existential*) if it can be axiomatized by a universal (existential) set of sentences.

A class of structures is said to be *arithmetical* (or *a variety*) if it is the class of all models of a theory (as always, in the lower predicate calculus). A class of structures is *universal* (*existential*) if it is the class of models of a universal (existential) theory.

Let M be a structure. The diagram of M, $D(M)$ is the set of all basic sentences which hold in M, for a given vocabulary which is supposed to include relation symbols for the relations of M, function symbols for the functions of M, where applicable, and terms for all individuals of M. We then say that the vocabulary is *adequate* for M. In particular if no function symbols and functions occur then the given vocabulary V is adequate for M if for every individual a of M there is an individual constant in V which denotes a. It is not difficult to see that M is a model of $D(M)$, for the given interpretation of the symbols of $D(M)$ in M. Conversely, if M' is a model of M then M' is isomorphic to an extension of M or, equivalently, M can be embedded in, or injected into, M'.

The embedding of one structure in another and similarly, the comparison of the properties of several structures presupposes that the relations and functions of these structures have been correlated in some way. This can be done conveniently by assuming that the relation symbols and function symbols of a given vocabulary V are interpreted in the several structures under discussion simultaneously and that the same applies to the individual constants of V if any. In practice, this is usually taken for granted: when we say that the field of real numbers, R, is an extension of the field of rational numbers, Q, we assume that addition in Q corresponds to addition in R, etc., and not that addition in Q corresponds to multiplication in R and vice versa. Thus, we assume implicitly, that the mappings from symbols of the language into entities of the structures under consideration commute with the mapping which expresses the injection of one structure into the other.

Let M and M' be two structures which are described by the sentences formulated in a given vocabulary, V. Then M is said to be *elementarily equivalent* to M' if for any one of these sentences, X, either X holds in M and M', or X holds neither in M nor in M' (so that $\neg X$ holds in both M and M'). The definition of elementary equivalence thus depends on the specified vocabulary, which however is frequently taken for granted.

If M' is an extension of the structure M, and M' is elementarily equivalent to M in a vocabulary V which is *adequate* for M (i.e., which includes terms as names for all the individuals of M, see above) then we say that M' is an *elementary extension* of M, in symbols $M \prec M'$ or $M' \succ M$. It is not difficult to see that this notion is independent of the particular choice of V.

The following example shows that two structures, M and M', where $M \subset M'$, may be elementarily equivalent for a particular vocabulary, without M' being an elementary extension of M. Let M' be a countable densely ordered set with first and last elements (e.g., the ordered set of natural numbers $0 \leqq x \leqq 1$). Choose an interior point x_0 of M' (e.g., $x_0 = 1/2$) and let M be the ordered substructure of M' which is given by the set $\{y \in M' \mid y \geqq x_0\}$. Then M is order-isomorphic to M' (e.g., in our particular example, by the mapping $x = 2y - 1$). It follows that if V contains just two relation symbols, for equality and order, then M' is elementarily equivalent to M with respect to V. However, M' is not an elementary extension of M for a sentence X which states that "x_0 is the first element" holds in M but not in M'.

There is some truth, and some exaggeration, in the assertion that all of (first order) model theory is based on the following result, which is known as the *finiteness principle* or *compactness theorem* of the lower predicate calculus.

THEOREM 2.1: *Let K be a set of sentences. If every finite subset of K possesses a model, then K possesses a model.*

The application of this theorem shortens numerous arguments in Algebra and therefore comes under the heading of "what every practitioner of Algebra ought to know." It is frequently useful in the following form.

THEOREM 2.2: *Suppose that the sentence X is a consequence of the set of sentences K. Then X is a consequence of a finite subset of K.*

The next section contains some simple direct applications of these theorems.

3. SOME FINITENESS ARGUMENTS IN ALGEBRA

Let K be a set of *axioms* for the notion of a commutative field. As indicated in section 1, and as can be verified without difficulty, these axioms can be expressed by sentences of the lower predicate calculus involving, in addition to the connectives, the variables, and the quantifiers, only the symbols $=$, $+$, $\cdot$, and 0. For example, the commutative law for addition can be written as

$$(\forall x)(\forall y)[x + y = y + x].$$

For any prime number p, we now wish to add a sentence X_p which expresses the assertion that the field to be described is of characteristic p. For this we choose

$$X_p\colon (\forall x)[\underbrace{(\cdots(x + x) + \cdots + x)}_{p \text{ times}} = 0]. \tag{3.1}$$

Thus, the set $K_p = K \cup \{X_p\}$ constitutes a set of axioms for the notion of a (commutative) field of characteristic p. At the same time, the set

$$K_0 = K \cup \{\neg X_2, \neg X_3, \neg X_5 \cdots, \neg X_p, \cdots\}$$

constitutes a set of axioms for the notion of a field of characteristic 0.

THEOREM 3.1: *Let X be a sentence of the lower predicate calculus expressed in the vocabulary of field theory ($=, +, \cdot, 0$ in addition to the usual logical symbols, as for K). Suppose that X holds for all fields of characteristic 0. Then X holds also for all fields of characteristic $p > p_0$ where p_0 is a positive integer which depends on X.*

Proof: The assumption of Theorem 3.1 is that $K_0 \vdash X$, X is a consequence of K. Hence, by Theorem 2.2 $\bar{K} \vdash X$, X is a consequence of $\bar{K}$ for some finite subset $\bar{K}$ of K_0. Since $\bar{K}$ is finite, there exists a set $K' = K \cup \{\neg X_2, \neg X_3, \cdots, \neg X_{p_0}\}$ such that $\bar{K} \subset K'$. Then $K' \vdash X$. But K' is satisfied by all fields of characteristic $p > p_0$ and so X is satisfied by all such fields also. This proves the theorem.

It is easy to give meaningful concrete examples to which this theorem applies. Thus, let $q_j(x_1, \cdots, x_n)$ $j = l, \cdots, k$, be a set of polynomials with integer coefficients and let $q(x_1, \cdots, x_n)$ be another polynomial with integer coefficients. The assertion that "*for all* $x_1, \cdots, x_n$, $q(x_1, \cdots, x_n) = 0$ *whenever* $q_1(x_1, \cdots, x_n) = 0$, $q_2(x_1, \cdots, x_n) = 0, \cdots, q_k(x_1, \cdots, x_n) = 0$" can be formulated in our vocabulary as a sentence X which is meaningful for all fields of characteristic p as well as for the fields of characteristic zero. It follows that if the assertion in quotes is true for all fields of characteristic zero, then there exists a positive p_0 such that the assertion is equally true in all fields of characteristic $p > p_0$. There is no essential difficulty in reaching this conclusion by means of ordinary algebra but a good deal of tedious work is avoided, here and in similar cases, by making use of 3.1.

Next we prove

THEOREM 3.2: *Let G be any (commutative or noncommutative) group. If every finitely generated subgroup of G can be ordered then G can be ordered (as a group).*

To have at least one fully worked example available, we first formulate a set of axioms K_Q for the notion of an ordered group. We choose, this time, to regard equality as an ordinary relation, $E(x, y)$ and to formulate K_Q in terms of it and of the two relations $Q(x, y)$ ("x is smaller than or equal to y") and $P(x, y, z)$ ("z is the product of x and y"). No individual constants will occur in K_Q.

$$(3.3)\quad (\forall x)E(x, x),$$
$$(\forall x)(\forall y)[E(x, y) \supset E(y, x)],$$
$$(\forall x)(\forall y)(\forall z)[E(x, y) \wedge E(y, z) \supset E(x, z)],$$
$$(\forall u)(\forall v)(\forall w)(\forall x)(\forall y)(\forall z)[[E(u, v) \wedge E(w, x) \wedge E(y, z) \wedge P(u, w, y)] \supset P(v, x, z)].$$

$$(3.4)\quad (\forall x)(\forall y)(\exists z)P(x, y, z),$$
$$(\forall w)(\forall x)(\forall y)(\forall z)[[P(w, x, y) \wedge P(w, x, z)] \supset E(y, z)],$$
$$(\forall u)(\forall v)(\forall w)(\forall x)(\forall y)(\forall z)[[P(u, v, w) \wedge P(w, x, y) \wedge P(v, x, z)] \supset P(u, z, y)].$$

$$(3.5)\quad (\forall x)(\forall y)[Q(x, y) \vee Q(y, x)],$$
$$(\forall x)(\forall y)(\forall z)[[Q(x, y) \wedge Q(y, z)] \supset Q(x, z)],$$
$$(\forall x)(\forall y)[Q(x, y) \wedge Q(y, x)] \supset E(x, y)].$$

$$(3.6)\quad (\forall u)(\forall v)(\forall w)(\forall x)(\forall y)(\forall z)[[P(u, v, w) \wedge P(x, y, z) \wedge Q(u, x) \wedge Q(v, y)] \supset Q(w, z)].$$

The sentences ("axioms") in the first group state that E is a relation of equivalence with substitutivity. (3.4) comprises the group axioms properly speaking. (3.5) states that $Q(x, y)$ defines a total order and (3.6) determines the connection between order and multiplication. The sentences of (3.3) and (3.4) alone constitute a set of axioms K_G for the concept of a group.

Now let G be any particular group, and let $D(G)$ be the diagram of

G. In order to show that G can be ordered as a group, we only have to prove that G can be embedded in an ordered group, for the restriction of the order relation in an ordered group to a subgroup yields an ordered group. In other words, we only have to show that $K_Q \cup D(G)$ is consistent, i.e., by Theorem 2.1, that every finite subset K' of $K_Q \cup D(G)$ possesses a model. Now K' can include only a finite number of individual constants, $a_1, \cdots, a_n$, say, denoting individuals $a'_1, \cdots, a'_n$ in G. Let G' be the subgroup of G which is generated by $a'_1, \cdots, a'_n$. G' is a model of $K' \cap D(G)$.

Suppose now that the assumptions of Theorem 3.2 are satisfied. Then we may equip G' with an order relation so that the resulting ordered group, G'_Q is a model of K_Q. But G'_Q is still a model of $K' \cap D(G)$ and so G'_Q is a model also of

$$K' = K' \cap (K_Q \cup D(G)) = (K' \cap K_Q) \cup (K' \cap D(G)).$$

This shows that K' is consistent and proves the theorem.

A very similar method yields proofs of embedding theorems. For example:

THEOREM 3.7: *Let I be a noncommutative integral domain. If every finitely generated subring of I can be embedded in a field then I also can be embedded in a field.*

We leave it to the reader to construct a proof.

4. WHAT IS AN ALGEBRAICALLY CLOSED FIELD?

From now on, whenever we discuss a class of structures, $\Sigma, \Sigma', \cdots$ we shall assume that the structures of the class are all *of the same type*. That is to say, the relations and functions (and, perhaps, some of the individuals) of the several structures can be identified by means of an underlying vocabulary. In this sense, we may, for example, talk of the class of all structures with equality and one binary operation, even though no axioms are specified for this class.

Let Σ be a class of structures. Σ will be called *existentially complete* if the following condition is satisfied.

(4.1) Let $M, M' \in \Sigma$ such that $M \subset M'$ and let X be any *existential* sentence formulated in a vocabulary V which is adequate for M. Then if M' satisfies $X(M' \models X)$, X is satisfied also by M.

Observe that the converse condition is satisfied automatically: if an existential sentence is satisfied by a structure M then it is satisfied also by all extensions of M.

The condition which is obtained from (4.1) by replacing the word *existential* in it by *primitive* (see section 2) is only apparently weaker than (4.1). For let X be an existential sentence formulated in the vocabulary V. Then X is logically equivalent to a sentence

$$(\exists x_1) \cdots (\exists x_n)[Q_1 \vee Q_2 \vee \cdots \vee Q_m],$$

where $Q_1, \cdots, Q_m$ are conjunctions of basic formulae. We may distribute the quantifiers across the disjunction, obtaining a sentence

$$X' = [(\exists x_1) \cdots (\exists x_n)Q_1] \vee [(\exists x_1) \cdots (\exists x_n)Q_2] \vee \cdots \vee [(\exists x_1) \cdots (\exists x_n)Q_m]$$

which is logically equivalent to X. Hence, if X holds in M', then one of the disjuncts, Y, of X' holds in M'. But each of these disjuncts is primitive so that if (4.1) is satisfied for primitive sentences then we may conclude that Y holds in M and so X' and X hold in M also. The advantage of this reduction is that it is frequently more convenient to check the truth of (4.1) for primitive sentences only.

Now let Σ_F be the class of commutative fields. Σ_F is arithmetical (a variety) for it is the class of models of a set of axioms $K = K_F$ for the notion of a commutative field. Let Σ_F' be the class of algebraically closed fields, $\Sigma_F' \subset \Sigma_F$. Then Σ_F' is again arithmetical, for by adding to K_F a sequence of sentences $Y_2, Y_3, Y_4, \cdots, Y_n, \cdots$, where Y_n states that "every monic polynomial of degree n has a root," we obtain a set of axioms K_C for the notion of an algebraically closed field.

It is one of the fundamental properties of any algebraically closed field M that if a finite set of polynomial equations and "inequations" with coefficients in M,

$$p_j(x_1, \cdots, x_n) = 0, \qquad q_j(x_1, \cdots, x_n) \neq 0 \tag{4.2}$$

possesses a solution in some (algebraically closed) extension of M

then it possesses a solution already in M. But the basic formulae which occur in a primitive sentence are just of the type (4.2) or are easily reducible to it (e.g., $p(x_1, \cdots, x_n) \neq q(x_1, \cdots, x_n)$ is equivalent to $p(x_1, \cdots, x_n) - q(x_1, \cdots, x_n) \neq 0$). We conclude that Σ'_F is existentially complete. Thus, as a first suggestion we might regard existential completeness as the "correct" metamathematical generalization of the notion of algebraic closedness. However, it will become apparent in the sequel that this place may also be assigned to a somewhat more sophisticated notion, which will now be introduced.

Let Σ be any class of structures. Σ will be called *model complete* if the following condition is satisfied:

(4.3) Let $M, M' \in \Sigma$ such that $M \subset M'$. Then $M \prec M'$, M' is an elementary extension of M.

Recalling the definition of an elementary extension which was given in section 2, we see that (4.3) is equivalent to the condition which is obtained from (4.1) by dropping the restriction that X be existential. Clearly then, if Σ satisfies (4.3) it satisfies also (4.1). This proves the "necessity" part of the following theorem.

THEOREM 4.4: *Let Σ be an arithmetical class. Then Σ is model complete iff it is existentially complete.*

To prove sufficiency, suppose that Σ is the class of models of a set of sentences K. Assume that Σ is existentially complete but not model complete. Then there exist triples $\langle M, M', X\rangle$ where M, $M' \in \Sigma$, $M \subset M'$, X is formulated in a vocabulary adequate for M, and X holds in M but not in M'. Since every sentence is logically equivalent to a sentence in prenex normal form in the same vocabulary, we may suppose that X has that form. Among all triples of this kind, we choose one for which the number of quantifiers in the prefix of X, k say, is as small as possible. Then $k > 0$, for if X is free of quantifiers and holds in M, it certainly holds also in M'.

We claim that X must begin with a universal quantifier. Suppose on the contrary that $X = (\exists y)Q(y)$. Since X holds in M, $Q(a)$ holds in M for some constant a. But then $Q(a)$ holds also in M', by the minimality of the number of quantifiers in X ($Q(a)$ contains only $k - 1$ quantifiers). Hence X also holds in M', contradicting the defining properties of our triples. Thus, let $X = (\forall y)Q(y)$ where $Q(y)$ may contain further quantifiers. By assumption $\neg X$ holds in M', i.e., $(\exists y)\neg Q(y)$ holds in M', i.e., $M' \models \neg Q(a)$ for some individual constant a. But as we transform $\neg Q(a)$ into prenex normal form in the usual way (by replacing existential by universal quantifiers, and by transferring the negation to the matrix) we obtain a sentence X' with $k - 1$ quantifiers only. It follows that X' holds in all extensions of M' which are models of K, and the same then applies to

$$\neg Q(a), \quad \text{to} \quad (\exists y)\neg Q(y), \quad \text{and to} \quad \neg X.$$

Now let $D(M')$ be the diagram of M'. Since $\neg X$ holds in all extensions of M' which are models of K, we have $K \cup D(M') \vdash \neg X$. Hence, by (2.1), there is a finite subset of $D(M')$, $\{Y_1, \cdots, Y_j\}$, say, such that $K \cup \{Y_1, \cdots, Y_j\} \vdash \neg X$. Thus, by the rules of the predicate calculus,

$$K \vdash Y_1 \wedge \cdots \wedge Y_j \supset \neg X.$$

Let $Y_1 \wedge \cdots \wedge Y_j = Y(a_1, \cdots, a_m)$ where we have displayed all the individual constants which denote individuals of $M' - M$ and, hence, do not occur in either K or $\neg X$. Then

$$K \vdash Y(a_1, \cdots, a_m) \supset \neg X$$

and, hence

$$K \vdash [(\exists z_1) \cdots (\exists z_m) Y(z_1, \cdots, z_m)] \supset \neg X.$$

Then the sentence $Z = (\exists z_1) \cdots (\exists z_n) Y(z_1, \cdots, z_n)$ is existential and even primitive, and has a meaning in M, and holds in M'. But Σ is existentially complete and so Z must hold also in M. But M is a model of K, and so $K \vdash Z \supset \neg X$ implies that M satisfies $\neg X$. This contradicts one of our assumptions on the triple $\langle M, M', X\rangle$ and proves Theorem 4.4.

As pointed out previously, the class of algebraically closed fields is existentially complete. Hence

THEOREM 4.5: *The class of algebraically closed fields is model complete.*

The algebraic resources involved in establishing the model completeness of an arithmetical class can be reduced in many cases by further metamathematical arguments. The long list of arithmetical classes that have been shown to be model complete includes —the class of densely ordered sets without first or last elements; the class of divisible ordered Abelian groups; the class of real closed fields (with or without order relations); the class of formally p-adic fields [1, 7].

A real closed ordered field can be described within the language of the lower predicate calculus as an ordered field in which every monic polynomial of odd degree has a root and every positive number has a square root. If no order relation is included from the outset the last condition is replaced by the requirement that for every $a \neq 0$ either a or $-a$, but not both, possesses a square root.

5. SUMS OF SQUARES

Accepting the fact that the class of real closed ordered fields is model complete, we shall now give a proof of Artin's theorem on positive definite polynomials with real coefficients (which represents a solution of Hilbert's seventeenth problem for this case).

Let M_0 be an ordered field and let M_1 be a (commutative, as always in this section) field which is an extension of M_0. Then it is obvious that in any ordering of M_1, extending the order of M_0, an element a which can be written in the form

$$a = c_1a_1^2 + \cdots + c_na_n^2, \qquad c_j \in M_0, \quad a_j \in M_1, \quad c_j \geqq 0, \quad n \geqq 1$$

must be nonnegative. Conversely, it is an elementary (though still remarkable) fact that if some $a \in M_1$ cannot be written in this form then there is an ordering of M_1 extending the order of M_0 for which a is negative.

Now let $p(x_1, \cdots, x_n)$ be a polynomial with coefficients in a real closed ordered field M_0. Suppose that $p(\xi_1, \cdots, \xi_n) \geqq 0$ for all $\xi_j \in M_0$. Then

THEOREM 5.1: *There exist rational functions $f_j(x_1, \cdots, x_n)$ with coefficients in M_0 such that*

$$p(x_1, \cdots, x_n) = \Sigma(f_j(x_1, \cdots, x_n))^2.$$

Proof: Evidently, it is sufficient to show that

$$p(x_1, \cdots, x_n) = \Sigma c_j(f_j(x_1, \cdots, x_n))^2$$

with f_j as indicated and $c_j \geqq 0$, $c_j \in M_0$ since any such c_j can be written as $c_j = d_j^2$, $d_j \in M_0$. Now if there is no such representation for $p(x_1, \cdots, x_n)$ then, taking $M_1 = M_0(x_1, \cdots, x_n)$ in the "elementary fact" quoted at the beginning of this section, there exists an ordering of M_1 for which $p(x_1, \cdots, x_n)$, as an element of M_1 is negative. That is to say, the sentence X which we may write briefly as

$$X = (\exists y_1) \cdots (\exists y_n)[p(y_1, \cdots, y_n) < 0] \tag{5.2}$$

holds in M_1. Now let M_2 be any real closed extension of M_1. Then (5.2) still holds in M_2. But the class of real closed fields is model complete and so (5.2) must hold also in M_0. This contradicts our assumptions and proves the theorem. Notice that for the above X, the assertion is true in M_1 precisely because the result of substituting the *elements* $x_1, \cdots, x_n$ of $M_0(x_1, \cdots, x_n)$ for the variables in p is just $p(x_1, \cdots, x_n) \in M_0(x_1, \cdots x_n)$.

Taking M_0 as the field of real numbers, we obtain Artin's theorem for this case. The corresponding theorem for polynomials with rational coefficients, also proved by Artin originally, can be established equally well by a slight modification of the above method.

S. Kochen has developed an important theory of rational functions with p-adic coefficients which is inspired by, and parallel to, the theory of positive definite polynomials given here.

6. COMPLETENESS

We have shown that the theory of algebraically closed fields is model complete. However, it is obvious that the theory is not complete in its own vocabulary. Indeed, since the class of models of K_C comprises *all* algebraically closed fields it cannot even decide the characteristic of a field, i.e., the sentences X_p (see 3.1) are neither deducible from nor refutable by K_C. Nevertheless, there is an important class of cases for which model completeness entails completeness. In order to introduce it we define the notion of a *prime model* of a class of structures Σ (for an underlying vocabulary V) as follows:

(6.1) $M_0 \in \Sigma$ is a prime model of Σ if for every $M \in \Sigma$ there exists an embedding $M_0 \xrightarrow{\phi} M$.

As explained earlier, we take it for granted that if an individual a in the underlying vocabulary denotes a_0 in M_0 then it denotes ϕa_0 in M.

For example, if Σ is the class of fields of characteristic $p \geqq 0$ then the prime field of that characteristic is a prime model for Σ. If Σ is the class of densely ordered infinite sets then any countable densely ordered infinite set without first or last element is a prime model for Σ but so is any densely ordered infinite set with first and last element. This shows that prime models are not necessarily unique up to isomorphism.

THEOREM 6.2: *Suppose that Σ is the class of all models of a set of sentences K, that Σ is model complete, and that it possesses a prime model. Then K is complete.*

Proof: Let X be a sentence formulated in the vocabulary of K, and let M_0 be a prime model in Σ. Then X either holds or does not hold in M_0 and we may assume, without loss of generality, that the first alternative applies. It then follows that X is a consequence of K, i.e., that X holds in all $M \in \Sigma$. For let $M_0 \xrightarrow{\phi} M$ be an embedding of M_0 into any given $M \in \Sigma$. Then $M_0 \models X$ implies that $M_1 \models X$ where M_1 is the ϕ-image of M_0. But then $M_1 \prec M$ and so $M \models X$ also. This proves the theorem.

Observe that if we restrict a model complete arithmetical class Σ by adding axioms (sentences) to a set K which determines Σ without enlarging the vocabulary, then the resulting class is still model complete. In particular, the class of algebraically closed fields of any fixed characteristic $p \geqq 0$ is model complete. Also, such a class includes a prime model—the field of absolutely algebraic numbers of the given characteristic. Hence

THEOREM 6.3: *The class of algebraically closed fields of given characteristic $p \geqq 0$ is complete.*

Thus, if X is a sentence in the language of field theory which is true in the field of complex numbers then (even if its truth has been established with the aid of topological concepts) it will be true in all algebraically closed fields of characteristic 0. This is a restricted, but nevertheless quite effective precise version of "Lefschetz' principle." For a considerably more flexible language, but with some restriction on the class of algebraically closed fields to be included in the conclusion, Lefschetz' principle was established recently by J. Barwise and P. Eklof [2].

7. AGAIN, WHAT IS AN ALGEBRAICALLY CLOSED FIELD?

We have seen that the subclass of algebraically closed fields of the class of commutative fields is existentially complete and model complete. Similarly, the subclass of real closed ordered fields of the class of ordered fields is existentially complete and model complete. The concepts of an algebraically closed field and of a real closed field were arrived at for good algebraic reasons, and the latter, which is due to Artin and Schreier, was to some extent inspired by the former. Within our present framework, it is natural to ask whether there exist some uniform formal conditions which single out the class of algebraically closed fields within the theory of commutative fields and the class of real closed ordered fields within the theory of ordered fields. We shall see that such a set of conditions does exist, in circumstances of striking generality.

Let Σ and Σ' be two classes of structures which are based on the same vocabulary. Σ' is said to be *model consistent* relative to Σ if every structure $M \in \Sigma$ can be embedded in a structure $M' \in \Sigma'$. For example, the class of commutative fields is model consistent relative to the class of commutative integral domains, but the class of fields (commutative or skew) is not model complete relative to the class of integral domains (commutative or not). Σ and Σ' are *mutually model consistent* if Σ' is model consistent relative to Σ and vice versa.

The notion of an elementary extension was introduced in section 2. It is not difficult to show that if $M_0 \prec M_1 \prec M_2 \prec \cdots$ is a chain of elementary extensions then the union (direct limit) $M = \bigcup_n M_n$

is an elementary extension of each M_n. We shall make use of this fact presently.

A class of structures Σ is *closed under elementary restriction* if for any $M \in \Sigma$, and for any structure M' such that $M' \prec M$, M' also belongs to Σ.

We now come to our crucial definition. For a given class of structures Σ, the class of structures Σ' (as usual, in the same vocabulary) is called *a model companion of* Σ if (i) Σ and Σ' are mutually model consistent, (ii) Σ' is model complete, and (iii) Σ' is closed under elementary restriction.

THEOREM 7.1: *A class of structures Σ possesses at most one model companion. That is to say, if Σ_1 and Σ_2 are model companions of Σ then $\Sigma_1 = \Sigma_2$.*

Proof: It is sufficient to show that $\Sigma_1 \subset \Sigma_2$; $\Sigma_2 \subset \Sigma_1$ then follows by symmetry. Let $M_0 \in \Sigma_1$. By condition (i), M_0 can be embedded in a structure $M_0' \in \Sigma_1$ and M_0' in turn can be embedded in some $M_1 \in \Sigma_2$. Next, M_1 can be embedded in some $M_1' \in \Sigma$, and M_1' in turn can be embedded in some $M_2 \in \Sigma_1$. Repeating the procedure, we obtain an extension M_3 of M_2 which belongs to Σ_2 and an extension M_4 of M_3 which belongs to Σ_1. Continuing in this way, we arrive at a chain of structures

$$M_0 \subset M_1 \subset M_2 \subset M_3 \subset M_4 \subset \cdots$$

such that the structures with even subscripts belong to Σ_1 and the structures with odd subscripts belong to Σ_2. But, by condition (ii) both Σ_1 and Σ_2 are model complete and so $M_0 \prec M_2 \prec M_4 \prec$ and $M_1 \prec M_3 \prec M_5 \prec \cdots$. It then follows from the above remark on chains of elementary extensions that the structure

$$M = \bigcup_n M_n = \bigcup_k M_{2k} = U \bigcup_k M_{2k+1}$$

is an elementary extension of all the M_n and, in particular $M_0 \prec M$ and $M_1 \prec M$. Now let X be any sentence formulated in a vocabulary which is adequate for M_0 and which holds in M_0. Then X holds also in M since $M \prec M$. But then X must hold also in M_1, otherwise $\neg X$ would hold in M_1 and hence, in M. We conclude that M_1 is an elementary extension of M_0. But then $M_0 \in \Sigma_2$, by condition (iii) above. This proves the theorem.

From now on, we may talk informally of *the* model companion of a class of structures, in the sense that there is *at most* one class of that kind.

Observe that a class Σ' which satisfies (ii) and (iii) is its own model companion. Hence, the model companion of the model companion of a class Σ is the model companion of Σ.

Referring back to section 4, we now see that the model complete theories discussed there did not, so to speak, arise by accident. Thus, the reader will check immediately that the class of algebraically closed fields is the model companion of the class of commutative fields, and the class of real closed ordered fields is the model companion of the class of ordered fields. Observe that in general a class Σ' will be the model companion of more than one class Σ. For instance, the class of algebraically closed fields is also the model companion of the class of commutative integral domains (as well as its own model companion).

There is a somewhat stronger notion than that of a model companion called a *model completion.* Thus, Σ' is called a model completion of Σ if (i) $\Sigma \supset \Sigma'$, (ii) Σ' is model consistent relative to Σ, (iii) Σ' is *model complete relative* to Σ, and (iv) Σ' is closed under elementary restriction. The last condition is satisfied automatically if Σ' is an arithmetical class. The notion introduced in condition (iii) means that if $M \in \Sigma$ and $M_1 \supset M$, $M_2 \supset M$, where M_1 and M_2 belong to Σ', then M_1 is elementarily equivalent to M_2 in a vocabulary which is adequate for M. It is not difficult to check that if Σ' is a model completion of Σ then it is a model companion of Σ. Hence, a class Σ has at most one model completion. On the other hand, it can be shown that a class may have a model companion which is not a model completion.

If Σ is an arithmetical class which is given by a set of axioms K, then we call the theory of the model companion (model completion) of K again its model companion (model completion) assuming the class in question exists. The following theorem which is given here without proof, (see [7], p. 128) settles the question of the existence of a model completion for a wide class of theories.

THEOREM 7.2: *Let K be a non-empty and consistent set of $\forall\exists$ sentences. That is to say, the sentences of K are in prenex normal form, and in each prefix no existential quantifier is followed by a universal quantifier. Then—in order that K possess a model completion K' it is necessary and sufficient that for every primitive formula $Q'(x_1, \cdots, x_n)$ in the vocabulary of K there exists an existential formula, $Q(x_1, \cdots, x_n)$, also in the vocabulary of K such that for any n individual constants denoting individuals of a model M of K, $Q(a_1, \cdots, a_n)$ holds in M iff $Q'(a_1, \cdots, a_n)$ holds in some extension of M which is a model of K.*

The proof of Theorem 7.2 actually yields a method for writing down a possible, though not necessarily the most convenient, set of axioms for K'. For example, for the notion of a differential field of characteristic 0, the conditions of Theorem 7.2 are satisfied and the model completion K' which is obtained in this way has been called the theory of *differentially closed fields*. It has been shown by Lenore Blum [3], using Morley's theory (see the end of the present paper) that the class of differentially closed fields which are extensions of a *given* differential field of characteristic 0 includes prime models, and recent work by S. Shelah shows that these are all isomorphic. However, according to the latest news they may possess proper endomorphisms over the groundfield, in which case they are not minimal.

For a given K, as in Theorem 7.2, suppose that K possesses a model completion, K'. Let $Q'(x_1, \cdots, x_n)$ be a primitive formula in the vocabulary of K and let $Q(x_1, \cdots, x_n)$ be a corresponding existential formula such as exists according to Theorem 7.2. Let M be a model of K and let $M' \supset M$ be a model of K'. Then we claim that for any $a_1, \cdots, a_n$ denoting individuals of M, $M \models Q(a_1, \cdots, a_n)$ iff $M' \models Q'(a_1, \cdots, a_n)$. Indeed, Theorem 7.2 shows directly that the condition is sufficient. To see that it is also necessary, suppose that $M \models Q(a_1, \cdots, a_n)$. By our assumption on Q and Q' there exists a structure $\bar{M}$ which is a model of K and an extension of M such that $\bar{M} \models Q'(a_1, \cdots, a_n)$. But Q' is existential and so it is easy to see that $Q'(a_1, \cdots, a_n)$ must hold also in all extensions of $\bar{M}$, and in particular in any $M' \supset \bar{M}$ which is a model of K'. But then $Q'(a_1, \cdots, a_n)$ holds in *all* extensions of M which are models of K', by condition (iii) in the definition of a model completion. This proves our assertion.

On the other hand, it can be shown that if K' is model complete, then *any* predicate $Q^*(x_1, \cdots, x_n)$ in the vocabulary of K' is equivalent with respect to K' to an existential predicate $Q'(x_1, \cdots, x_n)$, i.e.,

$$K' \vdash (\forall x_1) \cdots (\forall x_n)[Q^*(x_1, \cdots, x_n) \equiv Q'(x_1, \cdots, x_n)].$$

Since any existential predicate, in turn, is logically equivalent to a disjunction of primitive predicates, we finally conclude

THEOREM 7.3: *Let K' be the model completion of a set K. For any predicate $Q'(x_1, \cdots, x_n)$ (in the vocabulary of K or, which is the same, of K') there exists an existential predicate $Q(x_1, \cdots, x_n)$ (in the same vocabulary) such that for any $a_1, \cdots, a_n$ denoting elements of a model M of K, $Q(a_1, \cdots, a_n)$ holds in M iff $Q'(a_1, \cdots, a_n)$ holds in some extension of M which is a model of K', and hence, holds in all such extensions.*

For a given Q', a corresponding Q plays the role (and is in fact a generalization) of a "rational test" for the solvability of a system of equations in the theory of commutative fields such as is provided classically by a system of resultants, or for the existence of a root of an equation in an interval in the theory of ordered fields, as provided there by Sturm's chains. For suitable theories, e.g., for the two just mentioned, Q can be chosen so as to be free of quantifiers provided enough function symbols are introduced to ensure that the theory in question can be axiomatized by universal sentences only.

8. AGAIN, SUMS OF SQUARES

We shall now make use of Theorem 7.3 in order to establish the existence of a bound on the number of squares required to represent a positive definite polynomial.

Let M be an ordered field and let $p(x_1, \cdots, x_n)$ be a polynomial with coefficients in M such that $p(x_1, \cdots, x_n) \geqq 0$ for all $x_1, \cdots x_m$ in some—and hence in all—real closed extensions of M. Then the following assertion can be proved, by the method of section 5 or in some other way.

(8.1) There exist rational functions with coefficients in M, $f_j(x_1, \cdots, x_m)$, and elements c_j of M, $c_j \geqq 0$, $j = 1, \cdots, \lambda$ such that

$$p(x_1, \cdots, x_m) = \sum_{j=1}^{\lambda} c_j(f_j(x_1, \cdots, x_m))^2.$$

We are going to prove

THEOREM 8.2: *For any positive integer d there exists a positive integer B which depends only on m and d such that B is a bound on the number of required squares and on the degrees of the numerators and denominators of their bases, $f_j(x_1, \cdots, x_m)$, provided the degree of $p(x_1, \cdots, x_m)$ does not exceed d.*

Proof: Let $y_1, \cdots, y_n$ be the coefficients of the general polynomial of m variables and of degree d taken in some arbitrary but definite order. If we let this polynomial be the $p(x_1, \cdots, x_m)$ of (8.1) and if we impose a bound μ on the degrees of the numerators and denominators of the $f_j(x_1, \cdots, x_m)$, then the assertion of (8.1) can be expressed as a predicate $Q_{\lambda\mu}(y_1, \cdots, y_n)$ of the coefficients of $p(x_1, \cdots, x_m)$ (formulated, as usual within the lower predicate calculus). Let K be a set of axioms for the notion of an ordered commutative field and let $K' \supset K$ be a set of axioms for a real closed ordered field as considered previously. It is not difficult to see that, for $\lambda' > \lambda$ and $\mu' > \mu$,

(8.3) $$K \vdash (\forall y_1) \cdots (\forall y_n)[Q_{\lambda\mu}(y_1, \cdots, y_n) \supset Q_{\lambda'\mu'}(y_1, \cdots, y_n)].$$

On the other hand, let $Q'(y_1, \cdots, y_n)$ be a predicate which formally expresses the assertion that for all

$$x_1, \cdots, x_m, \qquad p(x_1, \cdots, x_m) \geqq 0.$$

Then the theorem on sums of squares states that for any $a_1, \cdots, a_n$ denoting elements of an ordered field M, $Q'(a_1, \cdots, a_n)$ holds in the real closed extensions of M iff at least one of the predicates

$$Q_{11}(a_1, \cdots, a_n), \qquad Q_{22}(a_1, \cdots, a_n), \cdots Q_{\lambda\lambda}(a_1, \cdots, a_n), \cdots$$

holds in M.

We now make use of Theorem 7.3 in order to obtain a predicate

$Q(y_1, \cdots, y_n)$ as described there. Thus, $Q(a_1, \cdots, a_n)$ holds in M iff at least one of the predicates

$$Q_{11}(a_1, \cdots, a_n), \cdots, Q_{\lambda\lambda}(a_1, \cdots, a_n), \cdots$$

holds in M.

It follows that the set of sentences

$$K \cup \{Q(a_1, \cdots, a_n), \neg Q_{11}(a_1, \cdots, a_n), \neg Q_{22}(a_1, \cdots, a_n), \cdots, \neg Q_{\lambda\lambda}(a_1, \cdots, a_n), \cdots\}$$

is not consistent, and hence that for some $\lambda = B$, the set

$$K \cup \{Q(a_1, \cdots, a_n), \neg Q_{11}(a_1, \cdots, a_n), \neg Q_{BB}(a_1, \cdots, a_n)\}$$

is not consistent. Hence

$$K \vdash Q(a_1, \cdots, a_n) \supset Q_{11}(a_1, \cdots, a_n) \vee \cdots \vee Q_{BB}(a_1, \cdots a_n).$$

Using Theorem 8.2, we deduce that

$$K \vdash Q(a_1, \cdots, a_n) \supset Q_{BB}(a_1, \cdots, a_n).$$

Now the only individual constant if any in K, is 0 and we may choose $a_1, \cdots, a_n$ different from 0. Then by the rules of the predicate calculus,

$$(8.4) \quad K \vdash (\forall y_1) \cdots (\forall y_n)[Q(y_1, \cdots, y_n) \supset Q_{BB}(y_1, \cdots, y_n)].$$

This shows that if $p(x_1, \cdots, x_m)$ is positive definite, i.e., if $Q'(y_1, \cdots, y_n)$ holds in all real closed extensions of M, then $Q_{BB}(y_1, \cdots, y_n)$ holds in M. In other words, the assertion of (8.1) is satisfied for $\lambda = B$, where B is also a bound on the degrees of the numerators and denominators of the $f_j(x_1, \cdots, x_m)$. The proof of Theorem 8.2 is complete.

For the case that the given field of coefficients of $p(x_1, \cdots, x_m)$ is the field of real numbers, or any other real closed field, an explicit bound B has been determined in recent years by A. Pfister [6]. Pfister shows that we may take $B = 2^m$ so that B is actually independent of the degree of $p(x_1, \cdots, x_m)$. On the other hand if we limit ourselves to the case that M is the field of rational numbers then there is still no purely algebraic method for dealing with the problem. In these cases we may of course omit the coefficients c_j in

(8.1) since every real number is a square and every positive rational number is the sum of four squares.

9. THE EXISTENCE OF MODEL COMPANIONS

Although we showed earlier that, quite generally, a class of models Σ cannot have more than one model companion Σ' our discussion of actual model companions so far has been confined to cases where both Σ and Σ' are arithmetical classes. However, recent work has shown that actually, a model companion Σ' exists for *any* arithmetical class Σ, even though Σ' itself need not be an arithmetical class. We proceed to discuss this matter, for the most part without giving proofs.

There is no essential loss of generality in assuming that Σ is a universal class. For suppose, to begin with, that Σ is the class of models of an arbitrary set of sentences K. Let $K_\forall$ be the set of all universal sentences which are formulated in the vocabulary of K and which are deducible from K, and let $\Sigma_\forall$ be the class of models of $K_\forall$. Clearly, $\Sigma_\forall \supset \Sigma$, and it is not difficult to see that $\Sigma_\forall$ is precisely the class of all substructures of models of Σ. (Here again, we take the vocabulary of K for granted. Thus, we call M' a substructure of $M \in \Sigma$ only if all the constants, function symbols and (of course) relation symbols which occur in K and $K_\forall$ denote entities of M'.) Then $\Sigma_\forall$ is mutually model consistent with Σ and so a model companion Σ' of $\Sigma_\forall$, which is mutually model consistent with $\Sigma_\forall$ is mutually model consistent with Σ also. It follows that Σ' is the model companion of Σ. Observe that Σ is closed not only under passage to substructures but also under passage to unions of chains of arbitrary length (or to inductive limits). Then

THEOREM 9.1: *Σ possesses a model companion, Σ', which is a subclass of $\Sigma_\forall$.*

Theorem 9.1 was first proved by an adaptation of the forcing notion (introduced by Paul J. Cohen to prove the independence of the continuum conjecture) to Model Theory [8, 9]. Two more proofs have been given by Ed Fisher and by Greg Cherlin, to whom I am indebted also for streamlining the terminology used in the present paper. For reasons which are explained by the connection with forcing, the elements of Σ' have been called the *generic structures* of Σ. It turns out that for certain important arithmetical classes Σ, the elements of Σ' are elementarily equivalent although they do not constitute an arithmetical class. Among such Σ are the class of groups and the class of division algebras of given characteristic.

This concludes our discussion. There are however several important developments which fall within the scope of the title of this paper and which have not been included here. Two of them deserve special mention. They are: the theory of Ax and Kochen [1] which settled a conjecture of Artin on p-adic fields, and the theory of M. Morley [5] which represents a far reaching generalization of the classical extension theory of commutative fields.

BIBLIOGRAPHY

1. Ax, J., and S. Kochen, "Diophantine problems over local fields, I, II," *Amer. J. Math.*, **87** (1965), 605–630, 631–648; III, *Ann. of Math.*, **83** (1966), 437–456.
2. Barwise, J., and P. Eklof, "Lefschetz's principle," *J. Algebra*, **13** (1969), 554–570.
3. Blum, L. C., "Generalized algebraic theories: a model theoretic approach," Ph.D. Dissertation, MIT, 1968.
4. Malcev, A. I., "Untersuchungen aus dem Gebiete der mathematischen Logik," *Mat. Sb.*, **5**, 1 (1936), 323–336.
5. Morley, M. D., "Categoricity in power," *Trans. Amer. Math. Soc.*, **114** (1965), 514–538.
6. Pfister, A., "Zur Darstellung definiter Funktionen als Summe von Quadraten," *Invent. Math.*, **4** (1967), 229–237.
7. Robinson, A., "Introduction to model theory and to the metamathematics of algebra," *Studies in Logic and the Foundations of Mathematics*, 1963.
8. ———, "Infinite forcing in model theory," *Proceedings of the second Scandinavian Logic Symposium*, Oslo 1970, (pub. 1971), pp. 317–340.

9. ———, "Forcing in model theory," *Proceedings of the Internat. Congress of Mathematicians*, Nice, 1970, (pub. 1971) vol. 1, pp. 245–250.

10. Tarski, A., and J. C. C. McKinsey, *A Decision Method for Elementary Algebra and Geometry*, 2nd ed. Berkeley, Los Angeles: 1951.

11. Tarski, A., and R. L. Vaught, "Arithmetical extensions of relational systems," *Compositio Math.*, **13** (1957), 81–102.

THE BEARING OF LARGE CARDINALS ON CONSTRUCTIBILITY*

Jack H. Silver

1. THE RUDIMENTS

Throughout this paper (except for metamathematical asides) we work in Zermelo-Fraenkel set theory with the axiom of choice (ZFC). Thus the only objects which figure in an official way in our deliberations are sets. However, it will prove expedient on a few occasions to speak of, say, the constructible universe or the universe of all sets. In all such cases, one can regard the suspect notion as a convenient way of speaking about a formula in the language of ZFC; accordingly all references to L (the name for the constructible universe) can be replaced in a fairly straightforward manner by references to the formula in one free variable which defines constructibility. Of course, the reader has the option of thinking in the framework of Gödel-Bernays set theory (with choice) which differs only inessentially from ZFC in that precisely the same formulas of the ZFC language

*This research was supported in part by NSF GP8746.

are theorems in the two theories, but in which one can treat L and related notions as classes.

Those readers who cherish metamathematical subtleties will perhaps find impertinent my inattentiveness to such matters as the difference between the language of set theory within which most of our discussion is carried on and the language which, properly speaking, is a set and which is used as one uses a language in model theory, certain structures being models of certain theories in the language, etc. Thus the language which figures in the definition of constructibility is a set and is a language in the second sense. It is really indifferent whether the σ_0 of section 2 is thought of as a sentence in the first or second sort of language, though our notation follows the second view, an approach which smooths the path slightly when one is, say, taking elementary substructures of models of σ_0. There is little point in laboring such language-level distinctions here; in axiomatic set theory it is vastly harder even than in other contexts to keep putting on the dog, metamathematically speaking. In certain kinds of arguments, only touched on here in section 4, such inattention to detail could prove utterly disastrous.

We turn now to several elementary notions and results in set theory. We call a set x *transitive* if it includes each of its members, i.e., if $y \in z \in x$, then $y \in x$. A binary relation R is said to be well founded if every non-empty set X contains an element which is R-minimal with respect to X, i.e., an element y such that zRy fails for all $z \in X$. The field of R is the union of the domain and range, i.e., $\{z\colon \text{for some } y, yRz \text{ or } zRy\}$. Those relations R are called *extensional* which satisfy this condition: if u, v are in the field of R, and for all z, zRu iff zRv, then $u = v$ (equivalently, if A is the field of R, then $\langle A, R\rangle$ satisfies the axiom of extensionality). If X is a set, let ε_x be the restriction of the membership relation to X, i.e., $\{\langle u, v\rangle\colon u, v \in X$ and $u \in v\}$; $\langle X, \varepsilon\rangle$ is short for the structure $\langle X, \varepsilon_x\rangle$. The Mostowski collapsing theorem, a stock-in-trade of professional set-theorists, asserts that if R is a well-founded extensional relation whose field is A, then $\langle A, R\rangle$ is isomorphic to $\langle M, \varepsilon\rangle$ where M is some transitive set. In saying that f is an isomorphism of $\langle A, R\rangle$ onto $\langle M, \varepsilon\rangle$, we mean simply that f is a 1–1 mapping of A onto M and that, for any $u, v \in A$, uRv iff $f(u) \in f(v)$. If f is such an isomorphism and M is transitive, it is easy to check that, for each $a \in A, f(a) = \{f(b): b \in A$ and $bRa\}$, in virtue of which f and M are uniquely determined. The Mostowski collapsing theorem can be proved with the help of this equation, using any one of several kinds of induction.

The notion of ordinal is so defined that each ordinal coincides with the set of (strictly) smaller ordinals. (More specifically, those sets which are transitive and are linearly ordered by the membership relation are called *ordinals*; this is referred to as the standard definition of 'ordinal.') Having assumed the axiom of choice, we stipulate that x is a *cardinal* iff x is an ordinal which cannot be put into a 1–1 correspondence with any smaller ordinal. A set X is said to have *cardinality* (or *power*) κ if κ is a cardinal which can be put into 1–1 correspondence with X. It cannot be too strongly insisted on that each cardinal κ is equal to the set of ordinals having power less than κ and that κ itself has power κ.

So far as model theory is concerned, we presuppose, in addition to the basic definitions, the (downward) Löwenheim-Skolem theorem and the basic model-theoretic result concerning ultraproducts. If $\mathfrak{A}$ is a structure and D is an ultrafilter on I, $\mathfrak{A}^I/D$ denotes the ultrapower of $\mathfrak{A}$ with index set I and ultrafilter D. Our conventions governing first-order languages stipulate that there is a fixed listing of variables $v_0, v_1, \cdots$ and that $\phi(a_0, \cdots, a_{n-1})$ is the result of substituting a_i for v_i, for each relevant i, in the formula ϕ. If $\mathfrak{A} = \langle A, \cdots \rangle$ and $\mathfrak{B} = \langle B, \cdots \rangle$ are structures appropriate to some first-order language $\mathscr{L}$ with equality, a mapping h is called an *elementary embedding* (or *elementary monomorphism*) of $\mathfrak{A}$ into $\mathfrak{B}$ if it is a 1–1 mapping of A onto B and if it preserves satisfaction, i.e., for any formula φ of $\mathfrak{B}$ and any list $a_0, \cdots, a_{n-1}$ of elements from A of suitable length,

$$\mathfrak{A} \models \varphi(a_0, \cdots, a_{n-1}) \quad \text{iff} \quad \mathfrak{B} \models \varphi(h(a_0), \cdots, h(a_{n-1})),$$

(where $\models$, as is customary, is the symbol for truth or satisfaction). In the special case that $\mathscr{L}$ is the language of set theory, whose predicate symbols are $=$ and ε, it will prove expeditious to have terms as well as formulas in the language. There is no loss of generality in supposing that each term τ has the form $\{v_1: \varphi\}$ where φ is a term-free formula of the language. In that case, if M is a set which contains the empty set and $a \in M$, then $\tau(a)^M$ is defined to be

$$\{b: \langle M, \varepsilon\rangle \models \varphi(a, b)\},$$

if that set is a member of M, otherwise 0. The L_α^M usage introduced in section 2 is governed by this convention, α playing the role of a.

2. THE FUNDAMENTALS OF THE CONSTRUCTIBLE UNIVERSE

Let x be a set. We say that a subset y of x is *describable* in x if there is a first-order formula φ (in the language with ε and $=$) which characterizes the elements of y in the structure $\langle x, \varepsilon\rangle$ in terms of certain (fixed) elements of x, i.e., for some $b_1, \cdots, b_n \in x$,

$$y = \{a: a \in x \quad \text{and} \quad \langle x, \varepsilon\rangle \models \varphi(a, b_1, \cdots, b_n)\}.$$

Suppose now that x is transitive, i.e., whenever $y \in z \in x$ it is the case that $y \in x$. Note that any member of x is itself a describable subset of x, owing to the transitivity of x; indeed, if $y \in x$, then

$$y = \{a: a \in x \quad \text{and} \quad \langle x, \varepsilon\rangle \models \varphi(a, b)\}$$

where φ is the formula $v_0 \in v_1$ and $b = y$.

Given this notion, the constructible universe, as we call it, can be built up quite simply in stages indexed by ordinals, using transfinite recursion. We define

$$L_0 = 0, \text{ the empty set},$$

$$L_{\alpha+1} = \text{the collection of subsets of } L_\alpha \text{ describable in } L_\alpha.$$

If α is a limit ordinal, $L_\alpha = \bigcup_{\beta<\alpha} L_\beta$.

L, the *constructible universe* itself, is the union of all the L_α. A set is called *constructible* if it is a member of L. If L_α is transitive, then every member of L_α is describable in L_α, or, in other words, is in $L_{\alpha+1}$; in consequence, $L_{\alpha+1}$, consisting as it does of all the members of the transitive set L_α and certain subsets of L_α, is itself transitive. That is the main step in the proof by transfinite induction that each L_α is transitive and that, for $\alpha < \beta$, $L_\alpha \subseteq L_\beta$. Moreover, observing first that the standard defining formula ψ in set theory of the notion "ordinal" is absolute in the sense that, for any transitive x, $\{a: a \in x$ and $\langle x, \varepsilon\rangle \models \psi(a)\}$ is precisely the set of ordinals in x, one can show quite readily by induction on α that L_α contains those ordinals which are less than α and no other ordinals. As might be expected, the members of $L_{\alpha+1} - L_\alpha$ are said to be constructed at stage α. Accordingly L_α consists of all sets constructed prior to stage α, and the unique ordinal constructed at stage α is α.

In respect to constructibility, Gödel [4] accomplished three great things. First, he invented it. Second, having defined the axiom of constructibility to be the sentence asserting that every set is constructible, he showed that all the axioms of ZFL, the theory whose axioms are the axioms of ZF together with the axioms of constructibility, are true in $\langle L, \varepsilon\rangle$, the constructible universe. This requires some elaboration, owing to the well-known impossibility of defining truth for some structures which are too large, e.g., for the universe of sets itself. If σ is a first-order sentence of set theory, let σ^L be the result of relativizing each of the quantifiers of σ to L, i.e., of replacing each $(\forall v_i)$ by $(\forall v_i \in L)$ and each $(\exists v_i)$ by $(\exists v_i \in L)$. Our claim can now be explicated in this way: If σ is the axiom of constructibility or any axiom of ZF set theory, then σ^L is provable in ZF. Third, Gödel proved in the theory ZFL the axiom of choice and the generalized continuum hypothesis (abbreviated GCH and meaning that $2^{\aleph_\alpha} = \aleph_{\alpha+1}$ for each ordinal α). We shall assume all of these but the very last one concerning the GCH, whose derivation in ZFL will not take us too far afield since it can serve to illustrate the value of what we call "the fundamental principle of constructibility."

FUNDAMENTAL PRINCIPLE: There is a first-order sentence σ_0 of set theory which is true in every structure $\langle L_\kappa, \varepsilon\rangle$ where κ is a cardinal and such that, whenever $\langle M, \varepsilon\rangle$ is a transitive model of σ_0 and α is an ordinal in M, $L_\alpha^M = L_\alpha$.

This asserts simply that the notion of being constructible before stage α is a fairly absolute notion, one whose absoluteness can be insured by requiring that M satisfy σ_0. The sentence σ_0 should be designed to insure the absoluteness of all the notions which figure in the sequence of definitions leading up to the definition of L_α (the notion of being a formula, being describable, etc.) and should guarantee the existence in M of the sequence $\langle L_\alpha : \alpha < \beta \rangle$ for each ordinal $\beta \in M$. The reader will profit from the discussion of related matters in Karp [6].

We now keep our promise to prove the GCH in ZFL. It is assumed that the axiom of choice has already been established in ZFL. An easy induction on α reveals that the cardinality of L_α is the same as the cardinality of α (i.e., the greatest cardinal less than α) except when α is finite, in which case we claim that L_α is also finite. The induction step is carried out by computing from the known cardinality of L_α the number of formulas with parameters from L_α substituted for variables, such as were used in the definition of "describable set." Given this fact, it will certainly suffice to see that

LEMMA (ZFL): *If κ is a cardinal and $S \subseteq \kappa$, then, for some $\alpha < \kappa^+$ (κ^+ = the successor cardinal of κ), $S \in L_\alpha$. Thus the power set of κ is included in L_{κ^+}.*[1]

Proof: By the axiom of constructibility, $S \in L_\beta$ for some ordinal β. Take λ to be a cardinal $> \beta$ for which $\langle L_\lambda, \varepsilon \rangle$ is a model of σ_0. Let $\langle N, \varepsilon \rangle$ be an elementary substructure of $\langle L_\lambda, \varepsilon \rangle$ which contains S, β, and every ordinal $< \kappa$ and which has cardinality κ[2]. Mostowski's collapsing theorem enables us to find an ε-isomorphism f mapping N onto some transitive set M. Since $\langle N, \varepsilon \rangle$ is an elementary substructure of the structure $\langle L_\lambda, \varepsilon \rangle$ in which the σ_0 of the Fundamental Principle holds, σ_0 will also hold in $\langle M, \varepsilon \rangle$. Moreover, if $y \in N$, then $f(y) = \{f(x) : x \in N \cap y\}$. Arguing by induction, we can thus conclude from the fact that $\kappa \subseteq N$ that f fixes each ordinal $< \kappa$ and that, as a consequence, $f(S) = S$. Owing to the absoluteness of L for L_λ (which is seen by appealing to the Fundamental Principle), we see that the assertion "$S \in L_\beta$" is true in $\langle L_\lambda, \varepsilon \rangle$, remains true in the elementary substructure $\langle N, \varepsilon \rangle$, and thence translates into the fact that "$f(S) \in L_{f(\beta)}$" is true in $\langle M, \varepsilon \rangle$. From the fact that $f(S) = S$ and the fact that M is a transitive model of σ_0, which guarantees the applicability of the Fundamental Principle to M, we conclude

1. The same proof shows that in ZFC all constructible subsets of κ are in L_{κ^+}.

2. This uses the Löwenheim-Skolem theorem.

at last that $S \in L_{f(\beta)}$. But, M being a transitive set of cardinality κ, $f(\beta)$, which is included in M, must indeed be an ordinal of cardinality at most κ, as desired.

It would hardly do to close this section without mentioning the profound inquiries into the constructible universe recently conducted by Jensen. He had first shown that Souslin's hypothesis is a theorem of ZFL. The following combinatorial principle, which is implicit in the subsequent proof of Solovay that Kurepa's conjecture is a theorem of ZFL (see [12]) was first expressly formulated by Jensen and myself and others (independently):

There is a function Q which assigns to each ordinal $\alpha < \omega_1$ a countable collection $Q(\alpha)$ of subsets of α such that the following holds: if X is any subset of ω_1, there is a closed cofinal subset C of ω_1 such that $X \cap \alpha$ and $C \cap \alpha$ are both members of $Q(\alpha)$ for each $\alpha \in C$.

This proposition, which is a theorem of ZFL, implies both Kurepa's conjecture and the negation of Souslin's hypothesis in ZFC.

3. LARGE CARDINALS

The interest in large cardinals, among which measurable cardinals figure most prominently, has its roots in at least three or four different lines of thought. In the first instance, mathematicians, having realized that there is no nontrivial translation-invariant countably-additive measure defined at every set of reals, were eager to learn whether this statement remains true if translation-invariance is dropped as a condition. For the latest information on this question, see Solovay [20]—it isn't stretching the truth too much to say that the answer is a qualified "no". Questions concerning real-valued measures naturally led people to inquire into measures which assume only the values 0 and 1, called 2-valued measures. Thus, at one time, such cardinals κ as admit a non-trivial (0 at singletons, 1 at κ itself) countably additive 2-valued measure defined at every subset of κ were termed measurable. This definition had the embarrassing defect that, under it, every cardinal greater than a measurable cardinal is itself measurable. In consequence, a more stringent definition has now been adopted, which replaces "countable-additivity" by "κ-additivity." (It is consoling to learn that the smallest cardinal measurable in the old sense is also measurable in the new sense.)

Yet another line of thought which has inspired large-cardinal experts is the attempt to generalize properties of ω. Inaccessibility can be regarded as such a generalization. Moreover, if by κ-additive we mean additive with respect to unions of length (strictly) less than κ, then ω admits an ω-additive nontrivial 2-valued measure on all subsets of ω. It is natural to ask whether any other cardinals κ have this property, that there is a κ-additive non-trivial 2-valued measure on all subsets of κ. This is precisely the definition of "measurable cardinal" we shall adopt. A host of other properties fall under this same rubric. Consider, for example, Ramsey's theorem: *If the unordered pairs from ω are partitioned into two pieces, there is a set X of cardinality ω all of whose pairs belong to the same piece of the partition.* Can we replace ω everywhere in this statement by κ and end up with a reasonable property? Such cardinals as possess this property and differ from ω are precisely the weakly compact, inaccessible cardinals (see [16]). Still another path to large cardinals leads from the effort to reflect properties of the whole class of ordinals down to cardinals. This approach, perhaps the most promising and powerful of all, lies outside the scope of this paper (see Reinhardt-Solovay [13]).

DEFINITION: Suppose κ is an infinite cardinal number. A collection D of subsets of κ is called a *non-trivial κ-complete ultrafilter on* κ if (1) every subset of κ or its complement is in D, (2) neither the empty set nor any singleton is in D, (3) if $\alpha < \kappa$ and, for each $\alpha < \kappa$, $X_\alpha \in D$, then $\bigcap_{\alpha<\gamma} X_\alpha \in D$. D is called a *normal ultrafilter* on κ if it is a non-trivial κ-complete ultrafilter and, for every function f whose domain is κ, if $\{\alpha\colon f(\alpha) < \alpha\} \in D$, then, for some γ, $\{\alpha\colon f(\alpha) = \gamma\} \in D$. Finally, a cardinal κ is said to be *measurable* if $\kappa \neq \omega$ and there is a non-trivial κ-complete ultrafilter on κ.

Notice that any statement about ultrafilters can be recast as a statement about 2-valued measures, simply by considering the measure which maps members of the ultrafilter into 1 and their complements into 0.

LEMMA [7]: *If κ is a measurable cardinal, then there is a normal ultrafilter on κ.*

Proof: Let D be any non-trivial κ-complete ultrafilter on κ. If f and g are functions from κ into κ, stipulate that $f < g$ iff $\{\alpha: f(\alpha) < g(\alpha)\} \in D$. It is evident that $<$ is a partial ordering, its transitivity following from the closure of D under finite intersections: if $f_1 < f_2$ and $f_2 < f_3$, then $\{\alpha: f_1(\alpha) < f_3(\alpha)\}$, including as it does the intersection

$$\{\alpha: f_1(\alpha) < f_2(\alpha)\} \cap \{\alpha: f_2(\alpha) < f_3(\alpha)\},$$

is itself in D. Moreover, there are no infinite descending chains under $<$, owing to the closure of D under countable intersections (recall that $\kappa > \omega$): Suppose $f_0 > f_1 > f_2 > \cdots$, i.e., for each $n \in \omega$, the set $\{\alpha: f_{n+1}(\alpha) < f_n(\alpha)\}$, which we call S_n, is in D. Thus the intersection $\bigcap_{n\in\omega} S_n$, being a member of D, is non-empty. If β is in this intersection, then $f_0(\beta) > f_1(\beta) > \cdots$, giving an infinite descending chain of ordinals, an impossible occurrence.

Consequently every non-empty set U of functions from κ into κ has a minimal element, for, if not, we could easily form a descending chain $f_0 > f_1 > f_2 > \cdots$ in U, at each step trading on the fact that f_n is not minimal in U. If $\alpha \in \kappa$, let $\hat{\alpha}$ be the constant function on κ whose value is always α. Set $U_0 = \{f\colon \text{for all } \alpha \in \kappa, \hat{\alpha} < f\}$. The identity function I on κ is a member of U_0 since, if $\beta \in \kappa$, $\{\alpha: I(\alpha) \geq \beta\} \supseteq \kappa - \beta$, a member of D. (That the complement of D in the power set of κ is closed under unions of length less than κ can be seen from the κ-completeness of D by passing to complements. Thus, each singleton being in the complement, no subset of cardinality $<\kappa$ is in D, whence each subset of κ whose complement has power $<\kappa$ is in D.)

Let h be a minimal element of U_0. As our candidate for a normal ultrafilter, we define $E = \{S: S \subseteq \kappa \text{ and } h^{-1}(S) \in D\}$ where $h^{-1}(S) = \{\alpha: h(\alpha) \in S\}$. Only the non-triviality and normality of E require comment; the κ-completeness of E comes from the fact that h^{-1} "preserves" intersections. As for non-triviality, note that $h^{-1}(\{\gamma\}) \in D$ would mean that h equals γ on a set in D, contrary to $h > \hat{\gamma}$. To check normality, suppose that f is a function on κ for which $\{\alpha: f(\alpha) < \alpha\} \in E$, i.e., $\{\beta: f(h(\beta)) < h(\beta)\} \in D$, i.e., $f \circ h < h$. By the minimality of h in U_0, $f \circ h \notin U_0$, so that for some $\delta \in \kappa$, it is not the case that $\hat{\delta} < f \circ h$, i.e., $\{\alpha: \delta < f(h(\alpha))\} \notin D$. This means that $\{\alpha: f(h(\alpha)) \leqq \delta\} \in D$, which, in view of the equation

$$\{\alpha: f(h(\alpha)) \leqq \delta\} = \bigcup_{\gamma \leqq \delta} \{\alpha: f(h(\alpha)) = \gamma\}$$

and the closure of the complement of D under unions of length $<\kappa$, implies at once that $\{\alpha: f(h(\alpha)) = \gamma\} \in D$ for some $\gamma \leqq \delta$. In virtue of the definition of E, this is to say that $\{\alpha: f(\alpha) = \gamma\} \in E$, just the conclusion needed for normality. Thus E is a nontrivial normal ultrafilter on κ, as desired.

It will be instructive briefly to reformulate the proof just completed in terms of an ultrapower. We form the ultrapower $\langle \kappa, < \rangle^{\kappa}/D$. The partial order $<$ defined above relates to this ultrapower in the following way: $f < g$ iff $f/D < g/D$ in the sense of the ultrapower, where f/D is the equivalence class of f. The $<$ of the ultrapower is easily seen to be a linear ordering, and, in the light of our discussion, must in fact be a well-ordering. h/D is the first element greater than all the elements $\hat{\alpha}/D$; indeed, one can show that $\{\hat{\alpha}/D: \alpha \in \kappa\}$ forms an initial segment and that accordingly h/D is the κth element. Thus, $f \circ h/D$, being less than h/D, must equal $\hat{\gamma}/D$ for some $\gamma \in \kappa$, as required.

If X is a set, let $[X]^n$ consist of all n-element subsets of X (here $n \in \omega$). Recall Ramsey's theorem in its most general form: if for some positive integers m and n, $[\omega]^n$ is partitioned into m pieces, say $[\omega]^n = I_1 \cup \cdots \cup I_m$, then there is such an infinite subset X of ω that $[X]^n \subseteq I_j$ for some j, i.e., all n-element subsets of X are in the same piece of the partition. This can be reformulated in terms of functions: if f is a function, from $[\omega]^n$ into m where m, n are positive integers, then there is an infinite subset X of ω such that f is constant on $[X]^n$. In the light of our remarks at the beginning of this section, it is perhaps not surprising that measurable cardinals should possess an analogue, indeed a very strong analogue, of the property embodied in Ramsey's theorem.

Rowbottom's Theorem [14]: *Suppose D is a normal ultrafilter on the measurable cardinal κ. If f maps $[\kappa]^n$ into λ where n is a positive integer and $\lambda < \kappa$, then there is an $X \in D$ such that f is constant on $[X]^n$.*

Proof: We proceed by induction on n. For $n = 1$, the statement says in essence that, if κ is divided into fewer than κ pieces, then one of the pieces is in D. But we have already said this in another guise, viz., the union of fewer than κ sets not in D is itself not in D.

Normality enters in at the induction step. Suppose the theorem is true for n. Suppose f maps $[\kappa]^{n+1}$ into some $\lambda < \kappa$. For each $\alpha \in \kappa$, consider the function f_α: $[\kappa]^n \to \lambda$ induced by f: for each $H \in [\kappa]^n$, let $f_\alpha(H) = f(H \cup \{\alpha\})$ where this is defined, 0 otherwise. By induction hypothesis, there is some $X_\alpha \in D$ for which f_α is constant on $[X_\alpha]^n$; the constant value assumed by f_α on $[X_\alpha]^n$ we call $g(\alpha)$. That g, a function from κ into λ, is constant on some $S \in D$, is just a variant of the case $n = 1$ above; denote the constant value γ. We now endeavor to find an $X \in D$ at each of whose $(n + 1)$-element subsets f assumes the value γ. Define

$$Y = \{\beta\colon \text{ for all } \alpha < \beta, \beta \in X_\alpha\}$$

and set $X = S \cap Y$. If $H = \{\alpha_1, \cdots, \alpha_{n+1}\} \in [X]^{n+1}$ where $\alpha_1 < \cdots < \alpha_{n+1}$, then, owing to $\alpha_2, \cdots, \alpha_{n+1}$ being in Y and, as a consequence, in X_{α_1}, it is the case that

$$f(H) = f_{\alpha_1}(\{\alpha_2, \cdots, \alpha_{n+1}\}) = \gamma,$$

the last equality because, α_1 lying in S, f_{α_1} equals γ everywhere on $[X_{\alpha_1}]^n$. It remains only to see that $Y \in D$, for then X, as the intersection of two sets in D, must itself be in D. But Y is what is called the diagonal intersection of the X_α. If $Y \notin D$, let $h(\beta) =$ first α such that $\beta \notin X_\alpha$, 0 if there is no such α. Since $\{\beta\colon h(\beta) < \beta\}$ includes $\kappa - Y$ which is in D (since $Y \notin D$), the normality of D implies that, for some $\delta < \kappa$, the set $\{\beta\colon h(\beta) = \delta\}$, which we call K, is in D. But, if $\beta \in K \cap (\kappa - Y)$, then $\beta \notin X_\delta$, in virtue of the definition of h. Thus, $K \cap (\kappa - Y) \cap X_\delta = \varnothing$, contrary to the fact that each of the three sets is in D.

DEFINITION: If X is a set, let $[X]^{<\aleph_0}$ be the collection of all finite subsets of X. If f is a function whose domain is $[\kappa]^{<\aleph_0}$ and $S \subseteq \kappa$, then S is called a set of *indiscernibles* for f if any two finite subsets of S having the same cardinality also have the same image under f. A cardinal κ is called a *Ramsey cardinal* if any function mapping $[\kappa]^{<\aleph_0}$ into some $\lambda < \kappa$ possesses a set S of indiscernibles having power κ [3].

As an easy corollary to Rowbottom's Theorem, we have

COROLLARY: *Any measurable cardinal is a Ramsey cardinal.*

Proof: Suppose κ is measurable and f maps $[\kappa]^{<\aleph_0}$ into some $\lambda < \kappa$. Let D be a normal ultrafilter on κ. For each n, denote by f_n the restriction of f to $[\kappa]^n$. In virtue of Rowbottom's Theorem, there is, for each n, an $X_n \in D$ such that f_n is constant on $[X_n]^n$. Patently $X = \bigcap_{n\in\omega} X_n$ is a set of indiscernibles for f having power κ. (This corollary was first proved by Erdos and Hajnal using another method.)

Finally, it is useful to remark that each Ramsey cardinal κ is inaccessible, that is to say, κ is not the limit of a sequence of ordinals having length less than κ, and, for each $\lambda < \kappa$, the power set of λ (i.e., the collection of all subsets of λ) has power less than κ. As for the first condition, if $\langle \eta_\alpha : \alpha < \lambda \rangle$ is an increasing sequence of ordinals of length $\lambda < \kappa$ tending to κ, then the function f which maps a pair $\{x, y\}$ from κ into 1 if, for some α, x and y are both between η_α and $\eta_{\alpha+1}$, 0 otherwise, clearly fails to have a set of indiscernibles of power κ (in this argument, as in the next one, we are using only a small part of the full Ramsey property, namely that which concerns functions on $[\kappa]^2$). Next suppose that λ is a cardinal $<\kappa$ whose power set has cardinality at least κ. That being so, there exists a function h mapping κ 1–1 into the set ${}^\lambda 2$ of all functions from λ into $2 = \{0, 1\}$ (noting that ${}^\lambda 2$ can be put into a natural 1–1 correspondence with the power set of λ). In this case, the function f which maps a pair $\{x, y\}$ (where $x \neq y$) into α if α is the least ordinal at which the functions $h(x)$ and $h(y)$ differ cannot have a set of indiscernibles containing even three elements.

4. CONSEQUENCES OF LARGE CARDINALS FOR THE CONSTRUCTIBLE UNIVERSE

Historically, Scott [15] was the first to establish a connection between large cardinals and constructibility, showing that, in ZFL, it is a theorem that there are no measurable cardinals. In other words, if there is a measurable cardinal, then there is a non-constructible set (a conclusion later improved by Rowbottom to: a non-constructible subset of ω). Indeed, that most conspicuous of all sets associated with a measurable cardinal, the (any) non-trivial κ-complete ultrafilter, turns out to be non-constructible.

Owing to the model-theoretic character of Scott's proof, it is incumbent on us to give some brief indication of (a variant of) this proof, though we shall rely in the proper and official discussion on very different methods which were discovered subsequently. Suppose κ is the smallest measurable cardinal and D is a non-trivial κ-complete ultrafilter on κ (it is D which is to be shown non-constructible). Suppose by way of contradiction, that D is constructible. Then, having seen that L is transitive, we can conclude that every member of D and hence every subset of κ is constructible (though ironically, as a later proof shows, even this is radically false). Let $M = L_\lambda$ where λ is a cardinal so large that $D \in L_\lambda$ and λ exceeds the successor cardinal of κ (indeed, with a little more sophistication, we could take M to be L itself).

Consider the ultrapower $\langle M, \varepsilon\rangle^\kappa/D$. This ultrapower is well-founded (or rather its ε relation is), in virtue of the argument given in the proof of section 3's lemma for the absence of infinite descending chains, and is moreover extensional, since it is elementarily equivalent to $\langle M, \varepsilon\rangle$. Invoking Mostowski's collapsing theorem, we obtain an isomorphism H of $\langle M, \varepsilon\rangle^\kappa/D$ onto some $\langle N, \varepsilon\rangle$ where N is a transitive set. Recall from the theory of ultraproducts that there is a canonical elementary embedding i of $\langle M, \varepsilon\rangle$ into the ultrapower $\langle M, \varepsilon\rangle^\kappa/D$ which maps each element into the corresponding constant function on κ. Composing i with H, we get an elementary embedding j of $\langle M, \varepsilon\rangle$ into $\langle N, \varepsilon\rangle$, which maps κ into an ordinal strictly greater than κ, because by arguments from the proof of section 3's lemma, the identity function on κ is an element of the ultrapower less than the constant function κ and yet greater than all the constant functions associated with smaller ordinals. From the fact that $\langle N, \varepsilon \rangle$ is elementarily equivalent to a model $\langle L_\lambda, \varepsilon \rangle$ of σ_0 and the axiom of constructibility, we may easily conclude that $N = L_\beta$ for some β. Since j is order-preserving, the order type of the set of ordinals in N, viz. β, must be at least λ. But λ was so chosen that $D \in L_\lambda$, *a fortiori* $D \in N$. A brief inspection of the property "being a non-trivial, κ-complete ultrafilter," whose only consequential quantifiers are universal ones, convinces us that

$\langle N, \varepsilon \rangle \models D$ is a non-trivial, κ-complete ultrafilter on κ,

and accordingly,

$\langle N, \varepsilon \rangle \models \kappa$ is measurable.

Thus $\langle N, \varepsilon \rangle \models (\exists v_0)(v_0$ is measurable and $v_0 < j(\kappa))$, since $\kappa < j(\kappa)$. Since j is an elementary embedding,

$\langle M, \varepsilon \rangle \models (\exists v_0)$ $(v_0$ is measurable and $v_0 < \kappa)$.

To see that this is contrary to the supposition that κ is the least measurable cardinal requires our remark that every subset of κ is constructible (and hence in M, since $\lambda \geqq \kappa^+$). For each ultrafilter on a cardinal λ smaller than κ, M therefore contains the wherewithal needed to block λ-completeness. We have indeed arrived at a contradiction.

It is instructive to consider the affinity of this proof with one of the standard proofs that the first inaccessible cardinal is not measurable [5]. One supposes to the contrary, taking D to be a non-trivial κ-complete ultrafilter on the first inaccessible κ. Let M consist of those sets hereditarily of power less than κ, i.e., $x \in M$ iff, whenever $y_1 \in y_2 \in \cdots \in y_n \in x$, then y_1 has power κ. An alternative characterization of M: M is transitive and, for all subsets X of M, $X \in M$ iff X has cardinality less than κ. (It is well known that $\langle M, \varepsilon \rangle$ is a model of ZF.) As before, we form the ultrapower $\langle M, \varepsilon \rangle^\kappa / D$, which is then seen to be isomorphic to some $\langle N, \varepsilon \rangle$ where N is transitive. The identity function inducing a large element in the ultrapower, we can infer that $\kappa \in N$. But inaccessibility is a property whose only consequential quantifiers are universal.

Hence

$$\langle N, \varepsilon \rangle \models \kappa \text{ is inaccessible},$$

and

$$\langle N, \varepsilon \rangle \models (\exists v_0)(v_0 \text{ is inaccessible}).$$

This last statement implies $\langle M, \varepsilon \rangle \models (\exists v_0)$ (v_0 is inaccessible) which runs counter to the assumption that κ is the smallest inaccessible (as an easy absoluteness argument proves).

We turn now to what is in essence an argument of Rowbottom [14], the upshot of which is that, if there is a Ramsey cardinal, then there are only countably many constructible real numbers (i.e., subsets of ω). This argument is presented in some detail. We begin with Rowbottom's crucial model-theoretic lemma, which has considerable intrinsic interest. (In trying to understand this lemma, it is helpful first to consider the special case where $A = \kappa$.)

LEMMA [14]: *Suppose κ is a Ramsey cardinal and $\mathfrak{A} = \langle A, U, R_i \rangle_{i\in\omega}$ is a relational structure for which $A \supseteq \kappa$ and the cardinality of U is less than κ. Then $\mathfrak{A}$ has an elementary substructure $\mathfrak{B} = \langle B, B \cap U, S_i \rangle_{i\in\omega}$ such that the cardinality of $B \cap \kappa$ is κ while $B \cap U$ is countable.*

Proof: We first propose to associate to each finite $H \subseteq \kappa$ a countable elementary substructure (or rather, properly speaking, the universe of such a substructure) $F(H)$ in such a way that F is an increasing function of H. This is done quite readily by induction on the size of H: If $F(H')$ has been so defined for each H' properly included in H, use the Löwenheim-Skolem theorem to obtain a countable elementary substructure of $\mathfrak{A}$ which includes all $F(H')$ where $H' \subset H$; call the universe of this structure $F(H)$.

For each finite $H \subseteq \kappa$, define $G(H) = F(H) \cap U$. Since κ is inaccessible, the power set of U (which power set includes the range of G) has power less than κ, in virtue of which we may apply to G the Ramsey property of κ (the definition of "Ramsey" notwithstanding, it is nonetheless clear from that definition that only the cardinality of the range matters, not the range's being an ordinal), thereby obtaining a set X of indiscernibles for G having power κ, so that any two finite subsets of X having the same size have the same image under G. Let $B = \bigcup \{F(H): H \text{ finite} \subseteq X\}$, and take $\mathfrak{B}$ to be the induced substructure $\langle B, B \cap U, R_i \mid B\rangle_{i\in\omega}$ of $\mathfrak{A}$. Taking H_n to consist of each n of the first n elements of X, if H is an n-element subset of X, then $F(H) \cap U \overset{\text{df}}{=} G(H) = G(H_n)$, owing to the indiscernibility of X for G. Thus, $B \cap U$, the union of all such sets $F(H) \cap U$, is the countable set $\bigcup_{n\in\omega} G(H_n)$. Also $B \cap \kappa$, including as it does the set X of power κ, must itself have power κ. It remains only to see that $\mathfrak{B}$ is an elementary substructure of $\mathfrak{A}$. It is a theorem of model theory that the union of a directed system $\mathscr{A}$ (i.e., if $\mathfrak{A}_1$, $\mathfrak{A}_2 \in \mathscr{A}$, there is $\mathfrak{A}_3$ such that $\mathfrak{A}_1 \subseteq \mathfrak{A}_3$, $\mathfrak{A}_2 \subseteq \mathfrak{A}_3$) of elementary substructures of a model $\mathfrak{A}$ is itself an elementary substructure of $\mathfrak{A}$, a theorem whose proof is a very slight generalization of the proof (by induction on formulas) of the much better-known subcase dealing with elementary towers. Manifestly the $F(H)$'s form such a directed system.

It would have been more natural to choose at the outset a fixed sequence of Skolem functions for $\mathfrak{A}$, then to define $F(H)$ as the closure of H under these functions. The last step in the proof would then have presented no difficulties. However, in deference to those who abhor Skolem functions, it seemed amusing to develop the version just presented.

THEOREM [14]: *If there is a Ramsey cardinal, then the set of constructible subsets of ω is countable (and, in consequence, there is a nonconstructible subset of ω).*

Proof: Let κ be a Ramsey cardinal, and let U be the set of constructible subsets of ω. The set U having power less than κ (owing to the inaccessibility of κ), we may apply the preceding lemma to the structure $\mathfrak{A} = \langle L_\kappa, U, \varepsilon\rangle$ so as to obtain an elementary substructure $\mathfrak{B} = \langle B, B \cap U, \varepsilon\rangle$ for which $B \cap \kappa$ has power κ and $B \cap U$ is countable. $\langle B, \varepsilon\rangle$ is a well-founded structure which is a model of σ_0 and the axiom of constructibility since $\langle L_\kappa, \varepsilon\rangle$ is. Therefore, the same must be true of the transitive structure $\langle M, \varepsilon\rangle$ (given by Mostowski's collapsing theorem) isomorphic to $\langle B, \varepsilon\rangle$. Accordingly, $M = L_\beta$ for some β, which can be seen to equal κ by noting that the order type of the set of ordinals in B is κ while the set of ordinals in L_β is precisely β.

If h is the isomorphism of $\langle B, \varepsilon\rangle$ onto $\langle M, \varepsilon\rangle$, let U' be the image of $B \cap U$ under h, i.e., $U' = \{h(y): y \in B \cap U\}$ (an easy argument would show $U' = B \cap U$). Let U be the predicate corresponding to U in the language of $\langle L_\kappa, U, \varepsilon\rangle$. Since $(\forall v_0)(v_0 \subseteq \omega \rightarrow Uv_0)$ is true in $\mathfrak{A}$, it is also true in $\langle M, U', \varepsilon\rangle$. This is to say that every subset of ω in $M = L_\kappa$ is in the set U'; which we know to be countable. Having shown in section 2 that every constructible subset of ω is in L_{ω_1} and a fortiori in L_κ, we see now that the set of constructible subsets of ω is the countable set U'.

It is an immediate consequence of this theorem that $\omega_1^L < \omega_1$ if there is a Ramsey cardinal. By ω_1^L we mean that ordinal α for which it is the case that $(\alpha = \omega_1)^L$, in other words, (α is the first uncountable cardinal)^L. Here the superscript L is being used in the same sense as in section 2, to indicate the result of relativizing quantifiers to L. Suspending for a moment our metamathematical scruples, we can instead say that ω_1^L is the ordinal α such that

$$\langle L, \varepsilon\rangle \models \alpha \text{ is the first uncountable cardinal.}$$

Since, in ZFL, one has proved that there are $\aleph_1$ subsets of ω, and since, as we have said, the constructible universe is a "model" of ZFL, there is in L a 1-1 map of ω_1^L onto that set S in L which is the power set of ω in the sense of L, i.e., which satisfies the condition

$$\langle L, \varepsilon\rangle \models S \text{ is the collection of all subsets of } \omega.$$

But one sees readily enough that this set S is precisely the set of all constructible subsets of ω. Since S is known to be countable under the hypothesis of the theorem, the same must be true of ω_1^L. It would be possible to carry this present line of argument further so as to prove that, if $\beta = \omega_1$, then

$$\langle L, \varepsilon\rangle \models \beta \text{ is inaccessible.}$$

It would be hard to overstate the extent to which this theorem represents an enormous improvement over Scott's original result (bear in mind that any measurable cardinal is Ramsey). Owing to the transitivity of L, it is clear, for example, that the power set of an infinite cardinal cannot be constructible if there is a Ramsey cardinal, and accordingly, under that hypothesis, no ultrafilter on an infinite cardinal can be constructible. Thus the non-trivial κ complete ultrafilters which bulked so large in Scott's treatment lose all their distinctiveness in this respect. (Note also the weakening of the hypothesis from "measurable" to "Ramsey." It can be shown [16] that the first measurable cardinal, if there is one, is much larger than the first Ramsey cardinal.)

Even the latest version we have presented is susceptible of drastic improvement. This is best accomplished by means of techniques discovered by Ehrenfeucht and Mostowski [2] and later exploited by Morley [10]. Owing to its intricacy, we can give only the barest outline of this procedure. See [16] for more details.

Suppose that κ is a Ramsey cardinal. The structure $\langle L_\kappa, \varepsilon\rangle$ can be endowed with Skolem functions in the following strong sense to form the structure $\mathfrak{A} = \langle L_\kappa, \varepsilon, f_\varphi\rangle_{\varphi\in\mathscr{F}}$ (where $\mathscr{F}$ is the set of formulas in the language appropriate to $\mathfrak{A}$): if φ is a formula in $\mathscr{F}$ the free variables of which are among $v_0, \cdots, v_n$ but in which v_n is free, then f_φ, the Skolem function corresponding to φ, satisfies, for all $a_1, \cdots, a_n \in L_\kappa$,

$$\mathfrak{A} \models \exists v_0\varphi(v_0, a_1, \cdots, a_n) \quad \text{iff} \quad \mathfrak{A} \models \varphi(f_\varphi(a_1, \cdots, a_n), a_1, \cdots, a_n)$$

(for example, one can arbitrarily fix a well-ordering of L_κ and take $f_\varphi(a_1, \cdots, a_n)$ to be the first element b such that $\mathfrak{A} \models \varphi(b, a_1, \cdots, a_n)$ if there is such an element, 0 otherwise). It should be realized that, in order to make correspond a Skolem function to each new formula which arises from the new Skolem functions previously introduced, it is necessary to carry out this process in ω stages.

An application of the Ramsey property strongly commends itself to us: we define a function g on $[\kappa]^{<\aleph_0}$ by the condition that, if $H = \{\alpha_0, \cdots, \alpha_{n-1}\} \in [\kappa]^{<\aleph_0}$ where $\alpha_0 < \cdots < \alpha_{n-1}$, then

$$g(H) = \{\varphi \in \mathscr{F} : \mathfrak{A} \models \varphi(\alpha_0, \cdots, \alpha_{n-1})\}$$

and, noting that the range of g has cardinality at most $2^{\aleph_0}$, consisting as it does of subsets of the countable set $\mathscr{F}$, and that $2^{\aleph_0} < \kappa$ since κ is Ramsey and hence inaccessible, we obtain a set of Z of κ indiscernibles for the function g, using the Ramsey property of κ. Thus, any two increasing sequences of elements of Z which have the same length satisfy precisely the same formulas in $\mathfrak{A}$, a situation which is sometimes described by saying that Z is a set of indiscernibles for the structure $\mathfrak{A}$. We choose X to be that set Z of κ indiscernibles which has the smallest ωth element among all such sets of κ indiscernibles. Finally, let Σ be the set of formulas satisfied by such increasing sequences from X, i.e., $\Sigma = \bigcup \{g(H) : H \in [X]^{<\aleph_0}\}$.

Now suppose that $\prec$ is an ordering of a set Y. From the arguments of Ehrenfeucht-Mostowski [2] and Morley [11] it is clear that there is a structure $\mathfrak{B} = \langle B, \varepsilon', g_\varphi \rangle_{\varphi \in \mathscr{F}}$ such that Y is a set of indiscernibles for $\mathfrak{B}$, any increasing sequence from which (i.e., Y) satisfies those formulas that are in Σ and that have at most such free variables as are accommodated by the sequence. In other words, if $z_0 < \cdots < z_{n-1}$ are in Y, then $\mathfrak{B} \models \varphi(z_0, \cdots, z_{n-1})$ iff $\varphi \in \Sigma$ and the free variables of φ are among $v_0, \cdots, v_{n-1}$. (It is possible to prove the existence of such a $\mathfrak{B}$ using the compactness or completeness theorem. Introducing an individual constant for each member of Y, write down the sentences describing the above situation and check that any finite subset of them has a model by referring back to X and $\mathfrak{A}$.) Finally, we may assume that B is generated by Y, i.e., B is the smallest subset of itself which includes Y and is closed under the functions g_φ (still another equivalent description: every element of B can be written in the form $\tau(z_0, \cdots, z_{n-1})^{\mathfrak{B}}$ where τ is some term formed from the function symbols in the language of $\mathfrak{A}$ and $\mathfrak{B}$ and $z_0, \cdots, z_{m-1} \in Y$, and $\tau(z_0, \cdots, z_{m-1})^{\mathfrak{B}}$ is the result of evaluating the term τ in the structure $\mathfrak{B}$ with z_i assigned to v_i for each i.) We can obtain this situation by cutting down the original $\mathfrak{B}$ to that substructure generated by Y, which must be an elementary substructure since the g_φ are Skolem functions just as the f_φ were.

Suppose further that $\prec$ is a well-ordering of Y which has order type λ where λ is some uncountable cardinal, i.e., the ordered structure $\langle Y, \prec \rangle$ is isomorphic to $\langle \lambda, \varepsilon \rangle$. We should like to see that $\mathfrak{B}$ is well founded (i.e., ε' is a well-founded relation), that $\langle B, \varepsilon' \rangle$ is in fact isomorphic to $\langle L_\lambda, \varepsilon \rangle$, and that, for any initial segment Y' of Y (under $\prec$) having no greatest element, that substructure of $\mathfrak{B}$ generated by Y' (using only the g_φ) is an initial segment of $\mathfrak{B}$, in a sense to be spelled out later. As we see later, it follows from these claims that, for every uncountable cardinal $\nu < \lambda$ $\langle L_\nu, \varepsilon \rangle$ is an elementary substructure of $\langle L_\lambda, \varepsilon \rangle$.

It is quite plausible that $\mathfrak{B}$ is well founded if $\lambda \leqq \kappa$, for in that case it is fairly clear that we can identify $\mathfrak{B}$ with the substructure of $\mathfrak{A}$ generated by the first λ elements of X (i.e., generated by means of the f_φ), since in each case Σ and λ completely determine the structure. To handle the general case, we suppose, by way of contradiction, that $\mathfrak{B}$ is not well founded, in which case there must be an infinite sequence z_n which satisfies the condition $z_{n+1}\ \varepsilon'\ z_n$ for each n. Manifestly, there is a countable subset C of Y which generates each z_n, that is, each z_n can be written in the form $\tau_n(c_1, \cdots, c_{m_n})$ where $c_1, \cdots, c_{m_n} \in C$. Let $\mathfrak{B}'$ be the substructure of $\mathfrak{B}$ generated by C, i.e., the universe of $\mathfrak{B}'$ consists precisely of those elements which can be represented in the form $\tau(c_0, \cdots, c_{n-1})^{\mathfrak{B}}$ where $c_0, \cdots, c_{n-1}) \in C$ and τ is a term.

If h is an ordering-preserving mapping of C into X, we claim that h can be extended to a monomorphism h' of $\mathfrak{B}'$ into $\mathfrak{A}$ given by

$$h'(\tau(c_0, \cdots, c_{n-1})^{\mathfrak{B}}) = \tau(h(c_0), \cdots, h(c_{n-1})^{\mathfrak{A}}).$$

It is in the first place necessary to see that this definition is unambiguous, in other words that different representations of the same element in the form $\tau(c_0, \cdots, c_{n-1})^{\mathfrak{B}}$ produce the same value on the right. By juggling variables, and remaining elements of C, we can reduce this to the claim:

If $c_0 \prec \cdots \prec c_{l-1}$ and $\tau(c_0, \cdots, c_{l-1})^{\mathfrak{B}} = \tau'(c_0, \cdots, c_{l-1})^{\mathfrak{B}}$, then $\tau(h(c_0), \cdots, h(c_{l-1}))^{\mathfrak{A}} = \tau'(h(c_0), \cdots, h(c_{l-1}))^{\mathfrak{A}}$. To say that the left-hand equation is true is to say that the sequence $\langle c_0, \cdots, c_{l-1}\rangle$ satisfies the formula $\tau(v_0, \cdots, v_{l-1}) = \tau'(v_0, \cdots, v_{l-1})$ in the structure $\mathfrak{B}$, which is the case iff that formula is in Σ, by virtue of the basic assumption on $\mathfrak{B}$ and the fact that $c_0 \prec \cdots \prec c_{l-1}$ are in Y. Then, recalling the definition of Σ, we have at once that the increasing sequence $\langle h(c_0), \cdots, h(c_{l-1})\rangle$ of elements from X satisfies that same formula in $\mathfrak{A}$, that is to say, the right-hand equation holds. Replacing $=$ everywhere by ε (or ε', as the case may be) in this argument gives us a proof that h' preserves ε. Hence $h'(z_{n+1}) \in h'(z_n)$ for each n, contrary to the axiom of regularity, which states that ε is well founded.

Having shown that $\mathfrak{B}$ is well founded, we can, in the light of Mostowski's collapsing theorem, find a structure isomorphic to $\mathfrak{B}$

whose universe is a transitive set and whose ε-relation is authentic set-membership. We may therefore assume without loss of generality that $\mathfrak{B}$ itself has these properties, i.e., B is a transitive set and ε' is the standard ε-relation on B. (Of course, the original set Y will in general have to be replaced by an isomorphic image.) $\langle B, \varepsilon\rangle$, as a structure elementarily equivalent to $\langle L_\kappa, \varepsilon\rangle$, is a model of σ_0 and the axiom of constructibility; accordingly, B must equal L_β for some ordinal β. Since B, including as it does Y, has cardinality at least κ, we shall have established that $\beta = \lambda$ once we see that every member of B has fewer than λ members (for, if $\beta > \kappa$, then L_κ, a set of cardinality λ, is a member of L_β).

We omit some of the details of this proof, details which can be found in [16] (or [17]). A rough description would run as follows: Let W be an initial segment of Y such that W has no greatest element. It will suffice to see that the substructure of $\mathfrak{B}$ generated W (using the g_φ) is a transitive set (for then, given any $x \in B$, say $x = \tau(y_0, \cdots, y_{n-1})^{\mathfrak{B}}$ where $y_0, \cdots, y_{n-1} \in Y$, take an initial segment W of Y which contains $y_0, \cdots, y_{n-1}$ and which has cardinality less than λ, and note that x is included in the set S of elements generated by W owing to the transitivity of S, which in addition is easily seen to have power less than λ). Suppose the transitivity claim fails; say $s = \tau_1(y_1, \cdots, y_n, z_1, \cdots, z_m)^{\mathfrak{B}} \in t = \tau_2(y_1, \cdots, y_n)^{\mathfrak{B}}$ where $y_1 < \cdots < y_n < z_1 < \cdots < z_m$ are in Y, $y_1, \cdots, y_n \in W$, and s is not generated by W. Let $z_1' < \cdots < z_m'$ be elements of W greater than y_n (recall that W has no greatest element). Since s is not generated by W, the element $s' \overset{\text{df}}{=} \tau_1(y_1, \cdots, y_n, z_1', \cdots, z_m')^{\mathfrak{B}}$ differs from s, in virtue of which the sequence

$$\langle y_1, \cdots, y_n, z_1', \cdots, z_m', z_1, \cdots, z_m\rangle$$

satisfies the formula φ defined as

$$\tau(v_0, \cdots, v_{n-1}, \quad v_n, \cdots, v_{n+m-1}) \neq \tau_1(v_0, \cdots, v_{n-1}, \\ v_{n+m}, \cdots, v_{n+2m-1}),$$

which is accordingly a member of Σ. That $s \in t$ translates into the fact that $\tau_1(v_0, \cdots, v_{n-1}, v_n, \cdots, v_{n+m-1}) \in \tau_2(v_0, \cdots, v_{n-1})$, which we call ψ, is a member of Σ. Now we refer these two observations back to $\mathfrak{A}$ and X. Choose $x_1 < \cdots < x_n$ in X and so choose increasing sequences $\langle z_1^\alpha, \cdots, z_m^\alpha \rangle$ for $\alpha < \kappa$ that $x < z_1^0 < z_m^\alpha < z_1^\beta$ whenever $\alpha < \beta$. Set

$$s_\alpha = \tau_1(x_1, \cdots, x_n, z_1^\alpha, \cdots, z_m^\alpha)^{\mathfrak{A}} \qquad \text{and} \qquad u = \tau_2(x_1, \cdots, x_n)^{\mathfrak{A}}.$$

Since φ is in Σ, $s_\alpha \neq s_\beta$ if $\alpha < \beta$, and, since ψ is in Σ, $s_\alpha \in u$. Thus u has cardinality at least κ, contrary to $u \in L_\kappa$.

Thus $B = L_\lambda$. We may turn our observation concerning initial segments to good account again. If ν is an uncountable cardinal less than λ, take W to be that initial segment consisting of the first ν elements of Y. If S is the set of elements generated by W, then S is a transitive set (as we have seen in the preceding paragraph), and is moreover isomorphic to L_ν (more precisely, $\langle S, \varepsilon \rangle$ is isomorphic to $\langle L_\nu, \varepsilon \rangle$) since S stands in the same relation to ν as B does to λ in the sense that the substructure of $\mathfrak{B}$ whose universe is S is generated by a set of indiscernibles whose increasing sequences satisfy the formulas in Σ. As a consequence of what we said in section 1, two different transitive sets cannot be isomorphic. Hence $S = L_\nu$. But S, closed as it is under the Skolem functions g_φ, is the universe of an elementary substructure of $\mathfrak{B}$. In particular, $\langle L_\nu, \varepsilon \rangle$ is an elementary substructure of $\langle L_\lambda, \varepsilon \rangle$.

It is also possible to show that Y is a closed cofinal subset of λ, that is to say, every element of λ is less than some element of Y, and further each limit, aside from λ itself, of elements of Y is in Y. We omit this proof altogether—suffice it to say that the proof makes strong use of our choice of X as the set of indiscernibles with the least possible ωth element.

At any rate, we have seen that (under our assumption that there is a Ramsey cardinal) the structures $\langle L_\lambda, \varepsilon \rangle$ as λ ranges over uncountable cardinals (i.e., uncountable in the real world and not just in L) form an elementary tower, i.e., whenever $\nu < \lambda$ are uncountable cardinals, $\langle L_\nu, \varepsilon \rangle$ is an elementary substructure of $\langle L_\lambda, \varepsilon \rangle$. In view of the well-known theorem concerning unions of elementary towers, it is tempting to assert that $\langle L, \varepsilon \rangle$ is an elementary extension of each $\langle L_\lambda, \varepsilon \rangle$. At first we shrink from this conclusion, owing to the difficulties in defining satisfaction and truth for large structures; upon further reflection, this dilemma is seen to suggest a way of defining truth for L (applicable only under some strong assumption, such as our present assumption that there is a Ramsey cardinal). If φ is a formula of set theory and $a_0, \cdots, a_{n-1}$ is a list of elements from L of suitable length, we say that $\varphi(a_0, \cdots, a_{n-1})$ is true in $\langle L, \varepsilon \rangle$ iff $\varphi(a_0, \cdots, a_{n-1})$ is true in $\langle L_\lambda, \varepsilon \rangle$ where λ is the least uncountable cardinal for which $a_0, \cdots, a_{n-1} \in L_\lambda$. One can check easily enough that this definition satisfies the inductive conditions one imposes on satisfaction, and that, under this dispensation, it is correct to claim that each $\langle L_\lambda, \varepsilon \rangle$, λ an uncountable cardinal, is an elementary substructure of $\langle L, \varepsilon \rangle$. Moreover, arguing from the claim that the set Y of our previous discussion was a closed, cofinal subset of λ, one can conclude there is a closed, cofinal class C of indiscernibles for $\langle L, \varepsilon \rangle$ such that every element of L is definable in $\langle L, \varepsilon \rangle$ from some finite list of members of C and every uncountable cardinal (of the real world) is in C. (Strictly speaking, the definability claim requires a more careful choice of the Skolem functions f_φ at the beginning of the proof—the f_φ should be so chosen that each one is definable by a formula in $\langle L_\kappa, \varepsilon \rangle$.) For more details, see my thesis [16] where the present argument first appeared. The set Σ, which can be redefined as the set of formulas satisfied in $\langle L, \varepsilon \rangle$ by increasing sequences of (real) uncountable cardinals has interesting metamathematical properties [16, 19] and was named $0^\#$ by Solovay who discovered those properties.

The conclusion that uncountable L_λ's form an elementary tower, a conclusion first inferred from the existence of a measurable cardinal by Gaifman using different methods, is even stronger than the previous theorem of this section, as we now indicate. Suppose $x \in L$ is an object first-order definable in $\langle L, \varepsilon \rangle$, that is, there is a first-order formula φ such that, for all $y \in L$,

$$\langle L, \varepsilon \rangle \models \varphi(y) \quad \text{iff} \quad y = x.$$

Since $\exists v_0 \varphi(v_0)$ is true in $\langle L, \varepsilon \rangle$, it is also true in $\langle L_{\omega_1}, \varepsilon \rangle$ an elementary substructure of $\langle L, \varepsilon \rangle$ by our assumptions. But, if $\langle L_{\omega_1}, \varepsilon \rangle \models \varphi(y)$, then, invoking that same assumption again, $\langle L, \varepsilon \rangle \models \varphi(y)$, whence $y = x$. Thus the element x is in L_{ω_1} and is in consequence a countable set (in the real world). In other words, every set first-order definable in $\langle L, \varepsilon \rangle$ is in L_{ω_1} and is countable. Certainly the constructible power set of ω, i.e., the set of constructible subsets of ω, is definable in L and is thus countable.

We now summarize the information gleaned from the indiscernibility argument, which reflects the efforts of Scott [15], Rowbottom [14], Gaifman, and Silver [16].

THEOREM: *If there is a Ramsey cardinal, then* (1) *for all* (*real*) *uncountable cardinals* $\nu < \lambda$, $\langle L_\nu, \varepsilon \rangle$ *is an elementary substructure of* $\langle L_\lambda, \varepsilon \rangle$, *and* (2) *there is a closed, cofinal class C of indiscernibles for* $\langle L, \varepsilon \rangle$ (*i.e., any two increasing sequences from C having the same length satisfy the same first-order formulas in* $\langle L, \varepsilon \rangle$) *which contains each* (*real*) *uncountable cardinal and for which it is the case that each element of L is first-order definable in* $\langle L, \varepsilon \rangle$ *from some finite list of elements in C.*

The conclusion, now often described in the literature as the existence of $0^{\#}$, is now known to follow from several mathematical assumptions, among which the axiom of determinateness and Chang's conjecture are perhaps best known (for the latter, see [8] and [18]). Kunen has also inferred our conclusion from the existence of a nontrivial elementary monomorphism from some $\langle L_\lambda, \varepsilon \rangle$ into itself, provided λ is a (real) cardinal (see [8] and also [18], where a simpler proof will be given).

BIBLIOGRAPHY

1. Cohen, P., *Set Theory and the Continuum Hypothesis.* New York: Benjamin, 1966.
2. Ehrenfeucht, A., and A. Mostowski, "Models of axiomatic theories admitting automorphisms," *Fund. Math.*, **43** (1956) 50–68.
3. Erdos, P., and R. Rado, "A partition calculus in set theory," *Bull. Amer. Math. Soc.*, **62** (1956), 427–489.
4. Gödel, K., *The Consistency of the Axiom of Choice and of the Generalized Continuum Hypothesis With the Axioms of Set Theory*, fourth printing. Princeton: 1958, 69 pp.

5. Hanf, W., Doctoral dissertation, University of Calif., Berkeley, 1962.

6. Karp, C., "A proof of the relative consistency of the continuum hypothesis," *Proceedings of the 1967 Congress of Logic, Methodology, and Philosophy of Science*. Amsterdam: 1969, 1–32.

7. Keisler, H. J., and A. Tarski, "From accessible to inaccessible cardinals," *Fund. Math.*, **53** (1964), 225–308.

8. Kunen, K., "Some applications of iterated ultrapowers in set theory," *Ann. of Math. Logic*, **1** (1970), 179–227.

9. Monk, J. D., *Introduction to Set Theory*. New York: McGraw-Hill, 1969.

10. Morley, M., "Categoricity in power," *Trans. Amer. Math. Soc.*, **114** (1965), 514–538.

11. ———, "Omitting classes of elements, the theory of models," *Proceedings of the 1963 International Symposium at Berkeley*, Amsterdam: 1965, 265–274.

12. Prikry, K., Doctoral dissertation, Univ. of Calif., Berkeley, 1968.

13. Reinhardt, W., and R. Solovay, to appear.

14. Rowbottom, F., Doctoral dissertation, University of Winconsin, Madison, 1964.

15. Scott, D., "Measurable cardinals and constructible sets," *Bull. Acad. Polon. Sci., Ser. Des. Sci. Math., Ast. et Phys.*, **9** (1961), 521–524.

16. Silver, J., Doctoral dissertation, Univ. of Calif., Berkeley, 1966 (has appeared in modified form in *Ann. of Math. Logic*, **3** (1971), 45–110).

17. ———, "Measurable cardinals and Δ_3^1 well-orderings," *Ann. of Math.*, **94** (1971), 414–446.

18. ———, "The consistency of Chang's conjecture," to appear.

19. Solovay, R., "A nonconstructible Δ_3^1 set of integers," *Trans. Amer. Math. Soc.*, **127** (1967), 58–75.

20. ———, "Real-valued measurable cardinals," *Proc.* 1967 *AMS Summer Set Theory Institute* (*UCLA*), to appear.

APPENDIX I—THE PREDICATE CALCULUS

A *relation structure* $\mathfrak{A}$ is a non-empty set A (the universe of $\mathfrak{A}$) with a set $\{R_i(i \in I)\}$ of relations R_i of finite degree on A (i.e., for each $i \in I$ there is a finite n such that $R_i \subseteq A^n$). We usually write $\mathfrak{A} = \langle A, R_i \rangle_{i \in I}$. Two relation structures have the same *similarity type* if their relation sets are indexed by the same set and corresponding relations have the same degree. If $\mathfrak{A}$ and $\mathfrak{B}$ have the same similarity type, then $\mathfrak{A}$ is said to be *isomorphic* to $\mathfrak{B}$ ($\mathfrak{A} \cong \mathfrak{B}$) if there is a 1-1 correspondence between the universes of $\mathfrak{A}$ and $\mathfrak{B}$ which preserves all the relations. Similarly, $\mathfrak{A}$ a *subsystem* of $\mathfrak{B}$ is defined in the natural way.

Sometimes the definition of a relation system is extended to include indexed sets of finitary operations (i.e., functions of A^n into A) and distinguished elements of the structure. Since an operation of degree n can be treated as a relation of degree $n + 1$ and a distinguished element $a \in A$ as a unary relation $\{a\}$, this is merely a matter of convenience. However, in this case, it is understood that a subsystem is closed under all operations and includes the distinguished elements.

Corresponding to each similarity type of structure there is an applied predicate language built up from the following symbols:

(i) a relation symbol of degree n for each relation of degree n of the structure,
(ii) a function symbol of degree n for each operation of degree n,
(iii) an individual constant symbol for each distinguished element,
(iv) a relation of degree 2, $=$ (to be interpreted as identity in every structure),
(v) an infinite set of variables $v_0, v_1, \cdots,$
(vi) logical connectives $\wedge$, $\vee$, $\neg$ (whose meanings are and, or, not, respectively) and a quantifier $\exists$ (read "there exists").

The set of *terms* is defined inductively by

(i) a variable is a term,
(ii) an individual constant is a term,
(iii) if f is a function symbol of degree n and $t_1, \cdots . t_n$ are terms, then $f(t_1, \cdots, t_n)$ is a term. (Formally, $f(t_1, \cdots, t_n)$ is the sequence $\langle f, t_1, \cdots, t_n \rangle$ but is written otherwise for appearance's sake. A similar remark applies to the definitions below.)

If R is a relation symbol of degree n and $t_1, \cdots, t_n$ are terms, then $R(t_1, \cdots, t_n)$ is an *atomic formula*. *Formulas* are defined inductively by:

(i) an atomic formula is a formula,
(ii) if φ and ψ are formulas, then the triples (see remark under (iii) above) $\varphi \wedge \psi$ and $\varphi \vee \psi$ and the pair $\neg\varphi$ are formulas,
(iii) if φ is a formula and v_i a variable, then $\exists v_i \varphi$ is a formula.

The derived expression $\varphi \supset \psi$ is interpreted as an abbreviation for $(\neg\varphi) \vee \psi$. Similarly, $\forall v_i \varphi$ is an abbreviation for $\neg\exists v_i \neg \varphi$.

An *interpretation* $\langle \mathfrak{A}, i \rangle$ of a predicate language is a relation structure $\mathfrak{A}$ (of the proper similarity type) together with a map i of the variables into A. The map i may be extended to all the terms of the language by defining i on an individual constant as the corresponding distinguished element of $\mathfrak{A}$ and $i(f(t_1, \cdots, t_n))$ as the result of applying the interpretation of the function f in $\mathfrak{A}$ to $i(t_1), \cdots, i(t_n)$.

The atomic formula $R(t_1, \cdots, t_n)$ is *valid* (for a given interpretation) if $\langle i(t_1), \cdots, i(t_n)\rangle$ is an element of the relation corresponding to R under the interpretation. Proceeding by induction, we may define the validity of every formula under a given interpretation.

We define inductively which variables *occur free* in a given term or expression:

(i) the only variable occurring free in v_i is v_i,
(ii) no variable occurs free in an individual constant,
(iii) a variable occurs free in a term $f(t_1, \cdots, t_n)$ or an atomic formula $R(t_1, \cdots, t_n)$ if it occurs free in one of the t_i,
(iv) a variable occurs free in $\varphi \wedge \psi$ or $\varphi \vee \psi$, if it occurs free in φ or ψ,
(v) a variable occurs free in $\neg\varphi$ if it occurs free in φ,
(vi) a variable occurs free in $\exists v_i \varphi$ if it occurs free in φ and is different from v_i.

An easy induction shows that the interpretation of a term or the validity of a formula is independent of $i(v)$ if v does not occur free in it. In particular, if a formula has no variables occurring free (it is then called a *sentence*), its validity depends only on the relation structure $\mathfrak{A}$ of the interpretation, but not on the assignment i of the variables.

If φ is a formula with its free occurring variables among $v_0, \cdots, v_{n-1}$, $\mathfrak{A}$ a relation structure and $a_0, \cdots, a_{n-1}$ elements of A, we say $a_0, \cdots, a_{n-1}$ *satisfies* φ in $\mathfrak{A}$ (written $\mathfrak{A} \models \varphi[a_0, \cdots, a_{n-1}]$) if φ is valid under an interpretation $\langle \mathfrak{A}, i\rangle$ with $i(v_j) = a_j$, $j = 0, \cdots, n-1$.

A subsystem $\mathfrak{B}$ of $\mathfrak{A}$ is an *elementary subsystem* of $\mathfrak{A}$ (written $\mathfrak{B} \prec \mathfrak{A}$) if for every formula φ and every n-tuple $\langle b_0, \cdots, b_{n-1}\rangle$ of elements of $\mathfrak{B}$, $\mathfrak{A} \models \varphi[b_0, \cdots, b_{n-1}]$ iff $\mathfrak{B} \models \varphi[b_0, \cdots, b_{n-1}]$. Two structures $\mathfrak{A}$ and $\mathfrak{B}$ of the same similarity type are *elementarily equivalent* (written $\mathfrak{A} \equiv \mathfrak{B}$) if, for every sentence φ, $\mathfrak{A} \models \varphi$ iff $\mathfrak{B} \models \varphi$. The reader should verify that $\mathfrak{A} \equiv \mathfrak{B}$ does not imply $\mathfrak{A} \cong \mathfrak{B}$ and that $\mathfrak{A} \subseteq \mathfrak{B}$ does not imply $\mathfrak{A} \prec \mathfrak{B}$.

Suppose $\mathfrak{A}_0 \subseteq \mathfrak{A}_1 \subseteq \cdots$ is an increasing chain (possibly transfinite) of relation structures. By $\bigcup \mathfrak{A}_i$ we mean the relation structure whose universe is the union of the universes A_i and whose relations $R_j (j \in J)$ are the union of the corresponding relations R_j of the structures $\mathfrak{A}_i$. If we replace $\subseteq$ by $\prec$ we obtain an *elementary chain.* The next theorem, which is due to Tarski, is very useful in the construction of models.

THEOREM: *If* $\mathfrak{A}_0 \prec \mathfrak{A}_1 \prec \cdots$ *is an elementary chain, then for each structure* $\mathfrak{A}_i$ *of the chain,* $\mathfrak{A}_i \prec \bigcup \mathfrak{A}_i$.

Proof: An easy induction on the length of the formulas.

A set of formulas is *consistent* if there is a single interpretation in which they are all valid.

The most important tool in model theory is the compactness theorem which follows.

THEOREM: *A set of formulas is consistent if every finite subset of it is consistent.*

The next most important theorem is the Löwenheim-Skolem theorem which says that a countable consistent set of sentences has a countable model. This has been generalized as follows:

THEOREM: *Let* $\mathfrak{A}$ *be a relation structure,* X *a subset of* A *and* κ *the cardinality of the language corresponding to the similarity type of* A. *Then*

(*i*) *there is* $\mathfrak{B} \prec \mathfrak{A}$ *such that* $X \subset B$ *and card* $(B) \leqq \kappa +$ *card* (X),
(*ii*) *if* $\mathfrak{A}$ *is infinite, then for every* $\lambda \geqq \kappa$, *there is* $\mathfrak{B} \succ \mathfrak{A}$ *with card* $(B) \geqq \lambda$.

Parts (i) and (ii) are often called the "downward" and "upward" Löwenheim-Skolem theorems respectively.

The *elementary diagram* of $\mathfrak{A}$ is the set of those sentences, of the language formed by adding a new constant for each element of $\mathfrak{A}$, which are valid in $\mathfrak{A}$. The *diagram* of $\mathfrak{A}$ is the set of those sentences of this enlarged language without quantifiers which are valid in $\mathfrak{A}$. It is obvious that $\mathfrak{A}$ is isomorphic to (an elementary) subsystem of $\mathfrak{B}$ iff $\mathfrak{B}$ satisfies the (elementary) diagram of $\mathfrak{A}$.

APPENDIX II—ULTRAFILTERS AND ULTRAPOWERS

Let X be a set $\neq \varnothing$. An *ultrafilter* on X is a maximal dual ideal in the Boolean algebra of all subsets of X. Stated another way, an ultrafilter is a set D of subsets of X such that:

(i) $\varnothing \notin D$,
(ii) $Y \subset X$ implies $Y \in D$ or $X - Y \in D$,
(iii) the intersection of any two elements of D is an element of D.

An ultrafilter is *principal* if it contains a smallest set. It can be shown that:

1. An ultrafilter is principal iff it contains a finite set, and
2. every infinite set has a non-principal ultrafilter on it.

Ultrafilters are used to define *ultraproducts* of relation structures. For simplicity, we shall discuss only ultrapowers (i.e., ultraproducts all of whose factors are equal) and relation structures $\mathfrak{A}$ which have relations but no functions or distinguished elements. Let I be a set $\neq \varnothing$, D an ultrafilter on I, and A^I the set of functions from I into A.

For $f, g \in A^I$, we define the equivalence relation $f \sim g$ to hold if the set $\{i \in I, f(i) = g(i)\} \in D$. Denote the equivalence class of f under $\sim$ by $\bar{f}$.

The relation structure $\mathfrak{A}^I/D$ (of the same similarity type as $\mathfrak{A}$) is defined as follows:

(i) the universe of $\mathfrak{A}^I/D$ is the set of equivalence classes $\bar{f}$,

(ii) the n-tuple $\langle \bar{f}_1, \cdots, \bar{f}_n \rangle$ satisfies R, R an n-ary relation iff the set

$$\{i \in I, \mathfrak{A} \models R(f_1(i), \cdots, f_n(i))\} \in D.$$

It must, of course, be verified that (i) and (ii) serve to define the relation structure $\mathfrak{A}^I/D$ (i.e., that the definition (ii) is independent of the representative of $\bar{f}$ chosen). The following theorem of Łoś is proved by induction on the length of formulas:

THEOREM: *If φ is a formula and $f_1, \cdots, f_n$ are elements of A^I, then $\mathfrak{A}^I/D \models \varphi[\bar{f}_1, \cdots, \bar{f}_n]$ iff the set $\{i \in I, \mathfrak{A} \models \varphi(f_1(i), \cdots, f_n(i))\} \in D$.*

Consider the map of $\mathfrak{A}$ into $\mathfrak{A}^I/D$ which carries each $a \in \mathfrak{A}$ into the constant function f with $f(i) = a$. It is a consequence of the preceding theorem that this map is *elementary* which means that it preserves all formulas. Identifying $\mathfrak{A}$ with its image under this map, we have $\mathfrak{A} \prec \mathfrak{A}^I/D$. The set $\mathfrak{A}^I/D - \mathfrak{A}$ is non-empty precisely when D is non-principal.

APPENDIX III—ZERMELO-FRAENKEL SET THEORY

The intuitive notions of naive set theory may be expressed as a set of axioms about a single binary relation $\in$. We describe one way to do this. It is convenient to make the following abbreviations:

$$x \subseteq y \quad \text{for} \quad \forall w(w \in x \supset w \in y),$$

$$y = \{w, v\} \quad \text{for} \quad \forall x(x \in y \leftrightarrow x = w \lor x = v),$$

$$\langle w, v\rangle \quad \text{for} \quad \{\{w\}, \{w, v\}\},$$

$$y = w \cup v \quad \text{for} \quad \forall x(x \in y \leftrightarrow x \in w \lor x \in v).$$

The axioms of Z.F. are:

1. Extensionality: $(x = y \leftrightarrow \forall z(z \in x \leftrightarrow z \in y))$.
2. Pairs: $\exists y(y = \{w, v\})$.
3. Union: $\exists y \forall x(x \in y \leftrightarrow \exists u(x \in u \land u \in z))$.
4. Power set: $\exists y \forall x(x \in y \leftrightarrow x \subseteq w)$.
5. Infinity: $\exists y \forall x((x \in y \supset x \cup \{x\} \in y) \land \exists x(x \in y))$.
6. Foundations: $(\exists x(x \in y) \supset \exists x(x \in y \land \forall z(z \in x \supset \neg z \in y)))$.

7. Replacement:

$$\forall w \forall x_1 \forall x_2 (\varphi(w, x_1) \cap \varphi(w, x_2) \supset x_1 = x_2)$$
$$\supset \forall z \exists y \forall x (x \in y \leftrightarrow \exists w (w \in z \land \varphi(w, x)))$$

for all formulas $\varphi(w, x)$ in which y does not appear free.

The axiom of replacement is actually an infinite set of axioms.

INDEX

INDEX OF SYMBOLS

(In the order in which they appear.)